Managing with Systems Thinking

Managing with Systems Thinking

Making dynamics work for you in business decision making

Michael Ballé

McGRAW-HILL BOOK COMPANY
London · New York · St Louis · San Francisco · Auckland
Bogotá · Caracas · Lisbon · Madrid · Mexico
Milan · Montreal · New Delhi · Panama · Paris · San Juan
São Paulo · Singapore · Sydney · Tokyo · Toronto

Published by
McGRAW-HILL Book Company Europe
Shoppenhangers Road, Maidenhead, Berkshire, SL6 2QL, England
Telephone 0628 23432
Fax 0628 770224

British Library Cataloguing in Publication Data

Ballé, Michael
    Managing with Systems Thinking: Making
    Dynamics Work for You in Business
    Decision Making
    I. Title
    658.4032

    ISBN 0-07-707951-5

Library of Congress Cataloging-in-Publication Data

Ballé, Michael
    Managing with systems thinking: making dynamics work for you in
    business decision making / Michael Ballé.
        p.   cm.
    Includes bibliographical references and index.
    ISBN 0-07-707951-5
    1. Management.   2. Decision making.   3 Critical thinking.
    4 Mental work.   5. Systems analysis.   I. Title.
    HD38.B235   1994
    658.4'03–dc20   94-7631

12345 CUP 97654

Typeset by BookEns Limited, Baldock, Herts.
and printed and bound in Great Britain at the University Press, Cambridge

For my parents, Catherine and Freddy

'Then there is an order in the world!'
I cried, triumphant.
'Then there is a bit of order in this poor
head of mine,' William answered.

The Name of the Rose
Umberto Eco

Contents

Preface

This book is for managers. This book is for anyone who feels that no matter how hard we push, 'the structure', or 'the system', or 'the organization' seems to push back harder; and for those of us who believe in trying but can see that sheer force is not always enough. The issues I have tried to address are particularly relevant to the running of projects within organizations. We try to understand how certain 'rational' behaviours create group dynamics that ripple throughout the organization. And, hence, how a project leader can sail his or her way through the firm's currents and tides to reach a safe harbour.

The book evolved out of numerous workshops, seminars and interviews with managers from various companies in which we faced real, nasty, messy business issues. In writing this book, I have tried to present a few concepts that can—hopefully—help foster insight in practical situations. It is not an academic work and does not pretend to either originality or exhaustivity—but to some solid common sense. The points that I make have, in a way, been dictated by practice more than by reason. They are the points that emerge, time and time again, in numerous working sessions and that seem to make sense to the participants. I believe that we cannot see something happening unless we already know what to look for. Columbus went looking for China, and found America instead. But centuries of sailors thought there was nothing there—so did not look. Over the years, we have found that too many of our working, practical problems are linked to the dynamic behaviour of organizations.

Most of us can deal intuitively with such dynamic behaviours as vicious cycles or virtuous circles, but often, intuition is not enough. I have tried to expose some building blocks that can help us deal with complex, interlinked dynamic situations by presenting them through examples. What I hope is that people confronted with these problems can then relate these concepts to their situation—and find workable solutions. The examples presented here are a patchwork of stories from the experience of managers I have been privileged to work with, from my own experience or from the literature. They have no other value than being a tool to illustrate some rather complex mechanisms. What I have attempted here is to show that we can deal with complexity simply—without being simplistic.

The systems thinking framework is particularly relevant to anyone involved with organizations, because organizational problems simply do not fit within simple linear thinking. Organizational issues are interlinked, dynamic and uncertain—in other words, complex. Traditional analysis, as in our habit of breaking everything down to its component parts and then studying the parts, is invaluable when dealing with machines or small groups, but will yield very few valuable insights to understand how our modern organizations behave. Furthermore, many of these organizations have been designed according to simplistic frameworks that do not represent reality with enough accuracy to make sensible choices.

As a result, 'problems' keep cropping up. Then, true to our philosophy, we spend a lot of time and effort 'problem-solving'. If we are lucky, the pain will go away and we can go back to our 'real work'. Unfortunately, this is not enough. For one thing, we usually attack the symptom rather than the root cause—because we do not have the intellectual tools to perceive the root cause. Our solutions are then likely to produce new 'problems' in the future rather than solving the ones we set out to deal with in the first place. In the meantime, the structural causes to our original concern are still there so that the problem will reappear for someone else to solve. Problem-solving is not enough as long as the thinking that created the problem is still active. The fundamental solutions lie in the way we design the structures and operational policies of our organizations. Policies must take into account effects on the organization as a whole rather than on the ailing operating part.

Systems thinking provides a vocabulary to describe and predict complex, dynamic mechanisms. What most experienced managers do anyway and call 'intuition' can now be formally used and shared among management teams in order to control the complex systems we create, rather than letting them control us. Systems behaviours can be counter-intuitive, yet they are none the less real. In this book, I have tried to capture a simple view of these complex occurrences—and have fun with it. I hope the result is a pleasant read.

Acknowledgements

Almost ten years ago, I stumbled upon a small book of collected essays of Gregory Bateson. After reading a few pages I felt as if somebody had finally found the light switch. *Steps to an Ecology of Mind*[1] made 'sense'. There was no real theoretical proposition in the book, no formalized doctrine, but a wealth of insights and observations about how the world works—or does not work. For once, these observations did not seem dry or abstract, but were consistent with how I personally experienced life. Bateson seemed to live in a universe where things were interconnected, dynamic, random and, yet, somehow ordered. Within complexity there is a pattern and he outlined the tools to draw it.

Being a self-taught 'systems thinker' I would like to thank authors whose remarkable works have paved the way for this book. I cannot quote here all the books that have inspired my work, but I would like to address particular thanks to Jay Forrester's *Industrial Dynamics*[2] and Peter Senge's *The Fifth Discipline*[3] from the MIT group, which have been particularly enlightening, also Barry Richmond's *Ithink User Guide*[4] and Donella Meadows's works.

From then on, I have pursued systems theory wherever I could find it and realized that a systemic vision of the world is emerging in fields as diverse as biology, cybernetics, social sciences, psychology, although there are, as yet, few applications to the field of management. When working as a systems dynamics modeller with Coopers & Lybrand, I realized that some of the mechanisms I was using to model market behaviour could be more directly used to shed some light on certain business problems. I followed the advice of my friend and colleague Doron Cohen, and wrote a short, friendly brochure on systems thinking for my then boss and mentor, John Rountree.[5] The success of this internal document led me to develop workshops with managers using these ideas. Most of what is in this book comes from experience of people in companies. I wish to thank all the people I have worked with, and in particular Michael Kightley and Carol Greenhalgh for their enthusiasm and insight in the early days of working with systems thinking in practice. All my thanks also to the rest of the team and in particular to Moira Evans and Rachel Parr.

I then worked with Thorogood Associates and developed more 'formal' systems thinking training courses. I would like to thank the team: Paul Balasky, Morag Banks, Patrick Dight, Jeremy Engasser, Kate Farmer, Mark Gibson, Julia Honnisberger, Trevor Jones, Tracy West and Rhona Worley for their warm support. At the end of one such session, Ian Dunham, one of the managers in the workshop, suggested that I write a short book summarizing basic systems thinking ideas and I would like to apologize for dismissing his idea at the time—it was great advice!

On a more personal front, I am very grateful to my wife, Lisa Ballé, for the patience and common sense that I have appreciated over years of 'systems thinking'. I would also like to thank Andrew Vincent, my father-in-law for long, fun and fruitful discussions about Life, the Universe and Everything. Many thanks to Andrew Gottshalk, Rob Abbott, Dennis Sherwood and Richard Hervey for their support in difficult times—as in good times. I am also very grateful to Mairi Mackinnon for correcting my grammar with great insight, and Marion Edsall for her skilled copyediting and Tessa Hanford for 'making it happen'. Finally, last but not least, I wish to thank my editors Kate Allen and Nathalie Burfitt for making this project happen and being wonderful listeners. As people say 'how can I know what I think if I don't hear myself say it?'; many thanks to all those who—willingly or not—have listened to me think.

Notes

1. G. Bateson, *Steps to an Ecology of Mind*, 1972, Ballantine Books.
2. J. Forrester, *Industrial Dynamics*, 1961, MIT Press.
3. P. Senge, *The Fifth Discipline: The Art and Practice of the Learning Organisation*, 1990, Bantam Doubleday.
4. Barry Richmond, *Ithink User Guide*, 1990, High Performance Systems.
5. M. Ballé, *Systems Thinking*, 1992, Coopers & Lybrand.

Introduction

Systems thinking has been around for almost a century, in many forms and disguises. It has been applied in various fields from cybernetics and computer sciences on the one hand to sociology and psychology on the other. S-curves have been used extensively in business since the 1960s and many economists have integrated aspects of systems theory in their analyses. Yet, as most fields increasingly adopt some concepts or techniques that stem from systems theory, the theory itself is still clouded in a shroud of esotericism and mystery. In the 1980s, there was a revival of interest in systems dynamics as a modelling technique for a range of issues, but yet again it has remained the domain of very highly specialized guru-type academics or consultants.

'Systems thinking' does not really exist as a unified field of research. It is at the intersection of many scientific disciplines and diversity is increased by the fact that the very term 'systems thinking' has different meaning for different groups of scientists. In the United Kingdom, critical systems theory as developed by P. Checkland, R. Flood, E. Carson and M. Jackson represents more a methodology—an approach—than specific theorizing. In the United States, systems thinking largely gravitates around the MIT and is solidly anchored in computer modelling (i.e. systems dynamics). This aspect of systems thinking has been successfully developed by J. Morecroft at the London Business School and M. Karsky in France. The French field of systems thinking is more centred on the role of the individual within the system than on overall systems behaviour. However, some seminal work has been done by thinkers such as R. Thom, E. Morin and J. de Rosnay. Within this diversity of works, I have centred my own understanding of 'systems thinking' around the feedback loop concept. Most of the other ideas should be represented here but the emphasis remains on representing structures of behaviour. Nothing in this book is new. The ideas have been around for a while. The author's main work and pleasure has been in the presentation of these concepts from a practical, workable point of view. The book provides a method, and specific, simple techniques to use systems thinking in practice. Systems thinking is not about solving corporate problems: it is about thinking.

Thinking is often taken for granted. The very logic of our thought processes is accepted as a 'given', something we cannot change. However, this 'logic' often goes wrong. Things just do not work out the way we figured they would. When that occurs, no matter how frequently, we tend to dismiss it and resign ourselves to the sad fact that 'accidents happen' and that the world is far too complex and weird a place for us to make any sense of it. Systems thinking is a practical way to challenge our old logic, to change ourselves in the very way we think and approach the world. It is very practical and down to earth—sometimes to a surprising extent in a world where accountants have taught us to think in terms of gross averages, percentages and balance sheets. This thinking is no less rational than what we commonly call 'rationality'. It is certainly more reasonable, and most people already use it and call it 'common sense'. Ultimately, what we wish to explore with you in this book is a way to capture common sense, and to use it in a more formal and systematic way. And, just like common sense, systems thinking often flies in the face of 'conventional wisdom' but that, of course, is also the fun of it.

Chapter 1

The problem

Managers are decision makers—or so the classical management theory would have us believe. It is true that managers occasionally have to make important decisions. These decisions will involve people, money and a lot of time and effort. However, these decisions can be difficult to make because such situations seldom occur. One might even say that organizations, as a rule, are built to avoid having people taking decisions except in certain well defined and controlled cases.

Of course, life is not like that, and more often than not, real decisions will have to be taken. But for most of us, managers or otherwise, decision making is not something that we indulge in very often. Most of the time, that is, every time we can, we just respond to situations in the way we always did. When the situation is new, or the stakes high, we recognize that we have a problem and feel that we have to 'make a decision'.

When that happens, we tend to be very confused, and seek a simple way to find 'the answer', or who to ask for 'the advice'. One of the reasons we have trouble dealing simply with the decision-making process is that our brain is a rather complex tool which will do a certain number of things simultaneously. To simplify to the extreme, in a given situation we observe, we react emotionally to what we have observed, we analyse, process and make judgements based on observations and feelings, and finally we act in order to make something happen. Because we are gifted creatures, we tend to confuse ourselves further by including the anticipated results of possible actions in the process. All this happens simultaneously, and over and over again till we reach the final state of tension and anguish that pushes us to do something to get it over and done with. People will make very different decisions according to when they reach that state.

Fortunately, most routine decisions have already been taken, either in the past or by somebody else, so our memory is triggered by the situation and we act accordingly. The main problem with this is that the situation has to change drastically for our memory not to jump to an 'I've seen this before' conclusion immediately. Sometimes we will do the wrong thing because the situation was not what it seemed. To use the parable of the boiled frog: if we put a frog in water, and slowly heat it, the frog will eventually let itself be boiled to death. In truth, the frog has no perceptual way to recognize the fact that the temperature is slowly increasing. We do. But are often misled by general confusion, so that we need an event, something we cannot dismiss outright, to catch our attention.

When this happens, we have to make a decision. If it is a decision where stakes are high and which involves others we need a 'decision-making process' to justify that decision: it was the best decision I could take at the time. The process might even help. As decision processes professionals, we strongly believe that it will; not so much because it will magically make everything all right, but because a process will ensure that the decision maker does not jump at the first available choice, but at least considers other alternatives. If the process is good, it will actually help generate alternatives— new possibilities that we had not thought of beforehand. In other words, it will foster creativity.

Yet, a fascinating aspect of decision-making processes is that they are not universal. They are linked to the epoch, culture, and even the kind of job or company with which we work. After trying emotional responses, honour, daring and many other action principles,[1] we have finally accepted that decisions should result in certain tangible benefits. This conclusion was reached, partly because an organization is built on the idea that individuals are replaceable, and partly because emotional decisions that would be favoured by the individuals progressively gave way to decisions from which the organization would materially benefit: in the crudest terms— profit. To paraphrase David Bohm, we have tried to replace the *rule of faith* by the *rule of reason* and ended up with the *rule of money*.[2]

In this context, most decision-making processes will involve some form of cost–benefit analysis. The very notion of acting to gain a profit implies that actions will be chosen through a certain number of trade-offs. This is very different from acting with honour, in which case the honour-bound course of action is often single and well defined. Somehow, acting for further benefit has become accepted as 'rational behaviour', and that has opened the scope for a rational approach of decision making. In the original sense of the term, rational often meant measurable, on the rather Victorian premiss that anything that is not measurable cannot really exist. Over the years, rationality has suffered many attacks and opened up slightly to non-measurable factors when not too inconvenient. Equally, we have made much progress into learning how to measure almost anything. The approach, however, stays the same.

It has flourished under several forms and names, with various degrees of complexity, and is often extremely valuable. We shall try to describe rational decision making in its simplest form. When considering a basic decision such as, 'shall I become a car mechanic?', the first step will be to list all factors involved in the decision, such as what my wife will say, how much money will I earn, how much do I know about cars and so forth; in doing this we break down the problem into its elementary parts. The second stage will then be to discriminate between 'cost' factors and 'benefit' factors. Thirdly, we will find a convenient way of weighting these factors against each other, and then make the decision that gets the best score. Interestingly, in this case, the decision is likely to be based on how one feels about becoming a car mechanic rather than the score.

In a personal decision, the 'gut feel' is often the winning criterion, but organizational decisions are too abstract, complex and accountable to allow 'gut feel' decision making. In fact, organizations have developed a splendid way of weighing factors: it is called accountancy, or finance. As the idea emerged that a company's goal was to make a profit, decisions could be evaluated against a simple criterion: 'will this course of action be profitable?'. We were then able to develop a host of indicators that can price almost anything in the firm in terms of money so that the question can ultimately be answered. Again, this wondrous idea has been pushed to the extreme and financial models of unheard complexity can now be developed.

The 'financial decision'

The 'financial decision' has gained increasing weight in companies as the calculation and data-gathering techniques have improved. As long as everything can be counted and trends can be projected, any decision can be treated in this way. Again, that has come to be accepted as the rational way to make decisions. It solves one major problem lurking in many firms: that of communication. Functional managers do not speak the same language. The everyday operating realities of sales, manufacturing, HR, and R&D environments are very different. Each area will have its own dialect and customs, judgements and values. As a consequence, a common language is needed to communicate across functions, and since everybody has to deal with budgets and is accountable for results, mostly in 'profit' terms, money is often chosen.

Although it is very good at 'keeping score' the financial language is very remote from operational problems of the different departments (in other words, how they work). A problem might appear on the financial analysis, but it will not give any direction how to solve it, nor, for that matter, what the problem actually is. The only message a financial analysis will convey is that certain departments perform well or badly. Although the more sophisticated 'variance analysis' attempts to explain why performance is good or bad in any department, it still rarely exposes the real root causes of problems. A financial decision is often reduced to an increase or decrease in the budget, trusting the operational managers to transform the money resource into operational problem solving. That is not always the result.

A finance person is likely to think in financial terms, he or she will think in terms of profit, cash, overheads and balance sheet. For example when confronted with the need to do something—to improve the productivity of a plant, say—a finance decision maker is likely to think through this accountancy mindset. Let us suppose that his or her staff come up with two projects: project A involves training the workers, setting up quality circles, making changes in the plant's operating rules and hiring a top consultant to supervise the operation; project B proposes investing the same amount of money to increase productivity by replacing groups of people by high-tech machines.

From a finance mindset point of view project A represents a risky cash outlay which is unlikely to increase this year's profit or the share value of the company. Project B, however, will enable the company to increase the balance sheet value by increasing its assets and to lower overheads by substituting machines for people. To a strictly finance person, project A does not make sense but project B sounds great. Hopefully, in real life, the process would not be as simple. The decision maker's opinion would be reinforced by his or her understanding of where productivity comes from (people or machines); of how workers could learn (can we teach these guys anything?); of what ripples the projects will create through the organization and which project his or her superiors would be likely to endorse. Of course, we can see that as a decision-making method, it is not good enough. But if many people in the company share the same mindset and it has been in place for a long time, then these conclusions are likely to be accepted as 'self-evident'.

In fact, finance based decision making is both more widely accepted and more frequently criticized as dangerous. Some awe inspiring examples are brought to light, such as the Vietnam War where American generals thought they were winning as long as the enemy body-count was higher than their own, the Detroit car industry that specialized in making money and forgot how to make cars and the overwhelming success of Japanese companies that seem to use such 'rational decisions' with moderation and make 'non-rational' choices.

The limits of classical rationality

As it happens, classical rationality has increasingly been contested. It had a serious setback in the 1970s at the hands of both academics and popular culture, but came back into fashion in the 1980s with a new name: limited rationality. In the 1990s, we do not talk about rational decisions any more, but about *best informed decisions*. The limitations of the 'rational' approach have been studied and highlighted in many fields. The most contested premise is that of perfect information. Classical decision theory[3] shows that a perfectly rational decision needs perfect information. Now, in decision theory, information has a cost; although a 'perfect' decision is theoretically possible, the cost of it would be prohibitive.

In the domain of the social sciences, rationality has been shown to be bounded, over and over again.[4] The obvious conclusion is that our brain is not a computer, human beings are not machines and, therefore, they cannot be rational. This is quite true. Interestingly, as the notion that rationality was limited gained acceptance, the 'financial decision making' gained popularity. Once we accept that financial decision making is not perfect, but we do not know any better method, short of going back to dark ancestral and savage ways, technology and particularly PCs and spreadsheets made the financial decision accessible to all parts of the firm. It is not restrained any more to major investment decisions but can be found almost anywhere as a legitimate way of making a decision. It is often justified by the feeling that we know it is not perfect, but nothing justified by the feeling that we know it is not perfect, but nothing else is and we have to do the best we can.

The truth is that the argument is perfectly valid, and that a 'rational' decision-making process is better than none. If we are able to measure costs and benefits accurately, there is no reason why we should not reach a correct solution that optimizes our profits. Unfortunately, there is a major assumption hidden in this reasoning: the notion that by increasing the parts individually, we will increase the sum of the parts. And there is the catch. Even assuming that all costs and benefits are measurable and comparable so that a judgement can be made on scores, one other major assumption remains. There will be no discontinuity in the way the costs and benefits evolve in the future: they are projectable. It is assumed that all things will continue to stay more or less in the same proportions.

This assumption is true only in steady conditions. When the environment is stable, with no major discontinuities, that assumption can be made. However, the tendency seems to be for increased chaos and disorder in recent years. Four mutually reinforcing trends—speed of communications, the development and use of new technologies, the growth of the service industry and the globalization of markets—have contributed to create a fast changing, complex and uncertain business environment. Any projection into the future of present behaviour is very likely to be wrong.

The second catch with 'rationality' occurs when from being an evaluation technique it is used by extension as action principles. The jump is easy to make: if everything can be measured in terms of costs and benefits, there is a simple way of increasing profits—reduce our costs and increase our benefits. Regrettably this attitude, which has resulted in numerous extravagant cost-cutting drives or sales campaigns, assumes that costs and benefits are independent: that it is possible to reduce a cost while maintaining benefits steady (and vice versa). Ironically, it is because it is possible to do so in the short term that this philosophy has been widely accepted. Nevertheless, the consequences appear soon enough, and in a chaotic way. An organization is a complex integrated system; to assume that we can treat certain parts of it in isolation is quite simply wrong. Whatever action we have, the effects will ripple with more or less intensity throughout the entire organization.

Although the general concepts of 'Total Quality Management' are now more or less accepted, many companies are still reluctant to put them into practice. One of the arguments put forward is the cost of quality: a quality process is going to be marginally more expensive than a traditional operation. However, if a quality programme is successfully implemented, the total costs which burden the company's operation will be drastically reduced: the cost of waste, the cost of losing customers because of defective products, the cost of product batches rejected at the quality inspection, etc. Yet these costs do not appear in most reporting systems. Consequently, they do not 'exist'. The cost of quality improvement, however, would be particularly visible (taking time off to think, to experiment, etc.) Although the idea of quality is more reasonable, from a cost–benefit point of view one trades a tangible, visible cost against a putative, pervasive but in the short term largely invisible benefit: no trade!

Rationality (or more accurately, what we commonly call rationality) is definitely bounded. At this present stage, 'rationality' is still very close to the idea of measurement. Yet several people have started to question whether our 'rationality' is itself logical. Logic itself is a construction built on abstract premises. In our operational use of rationality, these premises are accepted as facts. Yet are they facts, or little more than glorified assumptions?

The left-brain, right-brain debate

To look further into the problem of rationality, it is now necessary to find out more about how our minds work. It has been known for centuries that the brain is divided into a left and a right half, and that each side corresponds with the opposite side of the body. The ancient Egyptians discovered that injuries to one side of the brain caused a corresponding paralysis to the opposite side of the body; but it was only by the end of the nineteenth century that the question of the possible asymmetry of function of the two cerebral hemispheres began to be raised. As a result of a long history of research in that field, it is now generally accepted that the human cortex is formed of two complementary and anatomically slightly asymmetrical hemispheres: a right and a left half.

These two hemispheres are connected by a bundle of nerves—the 'corpus callosum'—which allows communication between the two halves of the brain. Since the pioneering work of Sperry at the end of the Second World War,[5] the functional specialization of each of the hemispheres has increasingly been studied. The results of these studies gave unequivocal evidence that each hemisphere possesses independent systems of perception. Sperry and Gazzaniga were furthermore able to prove that the two sides of the brain think in fundamentally different ways. For this amazing discovery, Sperry was awarded the Nobel Prize in 1981. Numerous studies on brain-damaged people support the findings of functional difference in the hemispheres. More evidence continues to be gathered, with the help of advanced technologies that can highlight which part of a person's brain is active when responding to different stimuli. As the results of these studies are being gathered, several dimensions are being suggested to describe the observed differences between the left and right brain, such as verbal versus non-verbal, propositional versus imaginative, analytic versus synthetic, explicit versus tacit, active versus receptive, intellectual versus emotional, and so forth. It has been shown for instance, that our left brain expresses its thoughts in words, while the right side tends to think directly in images. The left brain is also a one-step-at-a-time specialist: it thinks sequentially. On the contrary, the right brain processes information simultaneously and looks upon the situation as a whole, which explains its ability to think in images and to recognise and handle patterns. The left brain can be described as analytical whereas the right brain is holistic. Furthermore, intuition and emotion are said to be predominant right-brain functions, while the left brain seems to specialize in logic and reason. To summarize:

LEFT BRAIN	RIGHT BRAIN
Logic	Intuition
Reason	Emotion
Active	Receptive
Language	Visual recognition
Reading	Images, patterns
Writing	Depth
Linear processing	Parallel processing
Analysis	Synthesis
	Holism

Although some authors suggest that the functional differences between the left and right brain are absolute, in all likelihood the difference is relative. Each hemisphere is specialized in one type of processing but is able, at least minimally, to execute the other's function. Experiments confirm that it is only a preference of each half to process its own input first, only exchanging information with the other side once a considerable degree of processing has already been done. In any healthy person, both halves interact through the corpus callosum and the two types of consciousness are deeply linked.

This left-/right-brain distinction has been used as an analogy to describe the dominance of certain functions in human individuals or groups. Different types of personality, organizational functions and culture have been described in terms of left/right specialization. Mathematicians, for instance, will be described as 'left-brain personalities', while artists are more likely to be 'right-brain personalities'. Broadly, there are two complementary modes of thought: the first initially analyses and reduces information and events to bits and pieces; the second starts by relating, connecting and synthesizing information into wholes. Naturally, all cultures, national or corporate, use both left brain and right brain to a large degree, but research shows that some predominances do appear. In his cross-cultural comparative study 'The organization of meaning and the meaning of organization', researcher Fons Trompenaars argues that the left-brain/right-brain approach can be used to represent some more complex aspects of organizations:

> On the basis of these conclusions, hemispheric analogy can now also be applied to the conception of organizational structure which is, we repeat, a cultural construct and of the mind. However, [...] we stress the fact that it concerns ideal-typical description. The two ideal-typical conceptions of organizational structure, as well as the two ideal-typical sets of functions of the hemispheres, must be considered as analytical concepts. [...] In reality, however, there does not exist the 'right-brain' or the 'left-brain conception of structure', just as there are no right-brain or left-brain personalities or cultures. The strict distinction between these ideal-types, however, enables us to discern existing empirical tendencies toward, or preferences for, the one or the other. The actual conceptions actors have of structure, just as the actual functioning of the brain, may indeed involve both types of connotation. For the purpose of this study, we propose that one of the two connotations has primacy over the other for different actors and especially for different cultures.[6]

For instance, in our Western cultures there is a tendency to emphasize rational thought. The abilities to express oneself, read well and be analytical are highly valued whereas emotionality, intuition and spatial ability are less well regarded. To that extent, Western cultures are largely 'left brain'. In contrast, oriental cultures seem to allow a larger role to right-brain activities with a more holistic, timeless and synthetic approach. A similar right-brain predominance can be seen in Latin cultures where visual body language, facial expressions and variety in tone seem to have an important role in displaying open expressions of emotion.

Traditional left-brain approach

Our approach to rationality has traditionally been very left brain. Rational thought is supposed to be articulate and analytic. According to Charles Hampden-Turner, a senior researcher at the London Business School, this tendency evolved out of necessity:

> The English-speaking preference for individualism and analytical thinking gives advantages in the early stages of industrial development. Moreover, analysis is the best way of understanding simple mechanisms: it reveals their working by dissection when their function is essentially a sum of parts put to a purpose. But the advantage of this style of thinking diminishes when organizations grow large and products constitute complex wholes which encompass many satisfactions. Analysis is of less value when the focus shifts from machines to human systems and social teams because here nothing is gained from dividing the whole into parts. What a group knows and discovers is potentially more than can be carried away in the heads of its separate members.[7]

Such an approach to problems is both reductive into parts and sequential: first break down a problem into its elementary parts and then treat them one at a time. Our understanding of business is very time oriented, we are all acutely aware that 'time is money' and, furthermore, it tends to be discontinuous. We seldom see patterns, our attention is naturally focused on events. Conversations in organizations are dominated by events: last month's sales, new budget cuts, who just got promoted or fired, the new product our competitor just announced, and so on. This bias is strongly reinforced by the media who, from one day to the next, keep inundating us with newer, bigger, meaner events. Unfortunately, the world does not work solely in events, it evolves continuously. One of the greatest tragedies of our time is the wrecking extinction wave that we triggered: 100 species are said to disappear every day. Again, this is a left-brain formulation. Extinction, just like growth or differentiation, is not an event, it is a process. Our action, for instance, to save the whales will not be conclusive for decades. All we are doing is slowing down the extinction process.

Influence of left-brain preference on our decision making

Give me facts, not feelings
This left-brain bias, which we call 'rationality', strongly influences the outcomes of our decisions. It is not neutral, the very process of decision making has an impact on what kind of decisions are made. Typically, we have a long history of believing that hard, measurable facts are the only thing that count. This is reflected by operational structures in organizations where accountability is measured in terms of quantified goals to be achieved, such as 5 per cent increase in market share for marketing, or 10 per cent reduction in cycle time for production and so on.

This fact orientation often blinds us to non-measurable, but still immensely relevant forces. It took us almost a century to realize that staff morale was a necessary element of long-term corporate success. Many horror stories relate how people have 'intuitions' which are disregarded because they cannot be expressed with facts. Acting on such instincts would have avoided significant disasters. In certain extreme cases, such as the Challenger space shuttle accidents, facts are dismissed in the light of other facts: one engineer tried to convince the hierarchy that some pressure points would not hold in certain cold critical conditions, but his voice was lost in the maelstrom of 'more relevant' facts.

> On 28 January 1986, millions of viewers witnessed the tragic explosion of the Challenger space shuttle. Public opinion swung from astonishment to disbelief. This cannot be happening, could not have happened. We all know NASA's reputation for attention to security. It must have been a fluke, totally unpredictable technical failure. A catastrophe.
>
> Yet, in the weeks that followed the accident, it appeared that the accident was perfectly predictable. It was caused by a joint that was known to be likely to fail at low temperatures. Furthermore, technicians had tried to warn the flight directors of the risk involved in going ahead with a particularly cold morning take-off, but had been disregarded on account of 'keeping to schedule'. As it happens, NASA was under strong political and commercial pressure to increase the number of flights and this caused the warning to be ignored.
>
> In many ways, this tragic story symbolizes the risk of complexity. What seems to be a straightforward technical problem turns out to be a commercial–political–organizational–human problem. All these elements interlink to create real systems with real outcomes that we have the greatest difficulty in controlling.[8]

Individualism rather than group orientation

In addition, another consequence of our left brain bias is a strong preference for individual rather than group work. Although many organizations pay extensive lip-service to teamwork, very few management teams actually work as teams. Team spirit tends to be the first casualty of crisis; the moment things are not running smoothly, the groups tend to turn to a 'dog eat dog' behaviour. This can be explained by research that shows that in left-brain cultures the individual, rather than the team, will be held responsible for an incident of negligence at work. Furthermore, in left-brain terms, the poor job performance of an employee remains a criterion for dismissal regardless of age and previous record of working with the firm. Promotion will tend to be valued for the greater responsibility and higher income it provides to the individual rather than the opportunity of working with a new group of people.

> The main problem with individualism tends to be the 'They syndrome'; people will tend to identify with their position and not see how their actions affect others around them. Any problem is then 'their problem' or their fault, and much time and effort is spent on allocating blame or fighting for territory. This tends to lock everyone in a zero-sum game where what I gain is what we lose, and is contrary to overall gain of both we and me. For instance, in most companies the IT department tends to have a rather bad reputation. These 'techies' do not understand what the users want, do not even speak plain English, and have no idea of what the realities of business are. From the other side, users are seen to be fickle—changing their minds all the time—unrealistic and ungrateful. The truth is that both departments have to deal with different constraints. Users want very specific applications, and IT must maintain some consistency throughout the firm. Users are satisfied with the quick and dirty, but will blame IT every time a system falls over. At the end of the day, they both want the same thing: useful, running systems. Yet, each 'side' is always blaming the other for not behaving reasonably.

Focus on profits rather than well-being

Consistently with this approach, left-brain companies will tend to be focused on making profits rather than on the well-being of the members of the organization. A left-brain firm will see itself as a system designed to perform functions and tasks efficiently, with people hired to fit these tasks and functions. On the other hand, a right-brain company will believe that it is the quality of the relationships between its people that make these functions effective. For the right-brain organization, the well-being of the various stakeholders is the real goal rather than simply making a profit. In effect, the classical rational approach will tend to privilege the interests of shareholders and bankers. The concerns of the other members of the community are 'rationally' subordinated to these main interests. In a no less rational way, a classical Marxist approach will claim that only the workers' concerns are of any relevance. The point is, that in an organization, as in any community, the interests and concerns of all involved in any particular event have to be taken into account, not just those of a designated minority, whether shareholders, workers or executives.

In left-brain firms, this bias is mutually reinforcing with the financial approach to decision making, and in some cases, any other form of decision making is looked upon as suspicious, or plain silly. Yet, the emergence of Japanese and other East Asian companies and their much more right-brain approach to employment and manpower policies has finally cast a doubt on the ultimate efficiency of absolute focus on profits rather than people. This in no way has any 'moral' connotations, it does not mean that profits are 'wrong' and that only the interest of the people should prevail. A more holistic approach understands that both elements have to be somehow included in the equation. There is nothing wrong with making a profit (it is actually quite nice!), the slip occurs when every decision is considered in terms of how it is going to affect the 'bottom line'.

Technology rather than motivation
This bias in mindset tends to lead left-brain companies to prefer technology to people. Firstly, it makes sense from a financial point of view: by replacing a department by a machine we can increase our assets and lower our overheads. Secondly, machines are 'easier to deal with' than people. They do not complain, do not get bored (as far as we know) or have motivational problems. Thirdly, the all pervasive myth of 'human error' leads many managers to believe that a machine environment will be error-free. Author R. Pascale compares the attitudes of GM and Toyota as regards technology:

> By General Motors standards, the facility at Fremont (a Toyota run joint-venture plant) is an enigma. It is low-tech. It uses simple devices such as household-type clocks, to record downtime on the line, as contrasted to the $100 million information system that GM has been unsuccessfully experimenting with for the same purpose. Managers and workers move around the large plant on bicycles rather than on automated self-guided vehicles. Despite the fact that GM managers on the scene were submitting detailed, and increasingly emphatic, reports on Toyota's 'low-tech/high-motivation' formula, GM doggedly invested in technology ($60 billion over the period 1982–1985—

enough to buy Honda and Nissan). Why? Because the traditional rules at GM emphasize engineered solutions, not human or motivational solutions (except for a few cosmetic efforts at team building).[9]

Left-brain vision tends to be short term and task oriented

Similarly, because of its high awareness of time, and the very simple financial notion that a pound in pocket now is worth more than the same tomorrow (which ultimately leads to interest rates and the concept of discounting), left-brain scope tends to be fairly short term. This time orientation is reinforced by the sequentiality of left-brain thinking. Because actions are seen as one-step-at-a-time processes, time and planning become paramount. This in turn leads to task orientation: it is important to get things done without being certain of whether what we do is ultimately of value or not.

> Again, under the influence of the success of Japanese firms, we are now discovering the importance of process as opposed to individual tasks. Ironically, on the shop floor, most value-added tasks have been perfected over decades until very little productivity gain can be squeezed out of them, but they tend to constitute only about 5 per cent of the production activity. In many processes, 95 per cent of the activity is consumed by non-value-added work, such as moving things around, checking, sorting and so forth. It is in the processes themselves that real productivity gains can be found, but a traditional left-brain approach is blind to these opportunities because it is so focused on the tasks themselves.

Unbridled left-brain thinking can have disastrous results

Far too many organizational problems are born from unchecked left-brain thinking. The major problem with left brain thinking is that it will tend to focus on outputs rather than outcomes. Typically, any real-life situation is very complex and involves a multitude of interdependent parameters. An outcome is an overall result of our actions. Classical decision analysis theory very clearly shows that an optimization of all the parameters, one by one, cannot result in an optimal outcome. Yet our focus on outputs does just that, we try desperately to optimize outputs and end up with lousy outcomes. The left-brain difficulty in seeing a situation as a whole rather than as a sum of parts is directly responsible for our uneasiness with outcomes.

Moreover, the breaking down into parts approach, combined with a need for measurements, deceives us into dismissing important yet hard-to-measure parameters. These dismissed critical factors are often the cause of all the 'unexpected consequences' that can waste the best laid plans, sometimes to a tragic extent.

> One of the largest implementations of technology and rationality in modern times, the great Aswan dam in Egypt, built by Egypt's charismatic and progressive leader Nasser, is turning into a widespread environmental disaster that might ultimately result in a desertification of the world's oldest and most fertile valley. The fertile silts are held back behind the dam and do not enrich the lands by annual floods any more. As a result, Egyptian farmers have increasingly to use phosphates in order to maintain their level of production. The phosphates consistently exhaust the silt-poor soils which in turn increases their need for fertilizers. Furthermore, the lack of natural flood and drainage of arable lands increases the salinity of the soil. As sodium comes up to the surface, it kills the crops. If this ongoing dynamic is not changed, the fertile Nile valley has a fair chance of turning into a desert in a few decades. Ironically, the upper, previously arid, lands of Sudan, are becoming increasingly fertile as the dam keeps the silts upstream.

Such effects need not have such dire consequences, but in cases that are closer to home, left-brain thinking can lead us to kill the goose that lays the golden eggs: to make a short-term profit at a long-term cost. In many industries, investment and production are paramount to the success of companies, yet sales and finance are where most profits come from. In a 'success to the successful' dynamic, companies might be 'rationally' led to focus on the activities that generate immediate profits at the expense of those which ensure the very existence of the business. It has taken a long time for marketing to discover that profits do not come from sales, but from benefits to the customer. Yet that obvious statement is still widely ignored by companies which are too entrenched in traditional left-brain approaches.

Unsolved problems just come back

We naively believe that the future will be very much like the past, only bigger and better, and today's losers will catch up somehow if we just keep at it. As such, our policies and decisions usually treat the world as if it were divisible, separable, simple and infinite. The fact that the world is a complex, interconnected social–psychological–ecological–economic system is left for 'intellectuals' to ponder and write about. As a result, 'problems' keep cropping up and, true to our philosophy, we then spend a lot of time and effort 'problem solving'. Yet this is not enough.

We usually attack the symptom rather than the root cause, because we do not have the intellectual tools to *see* the root cause—we do not see something until we have the right metaphor to allow us to perceive it. Then, our 'solutions' are as likely as anything else to produce new problems in the future rather than solving the ones we set out to solve in the first place.

Furthermore, by attacking the symptom, the structural causes to the problem are usually still there, so that it will reappear for someone else to solve. This can be described as the 'garbage can' theory[10] of problem solving. A problem is seen as a garbage can. The can lands on someone's desk who will then chuck whatever solution he or she has available into the can. If the person is an accountant, he or she is likely to cut costs; if in marketing, to increase advertising spend, and so on. If one solution works, the can is emptied; but more likely it does not work. The person just kicks the can on to his neighbour's desk, who will then try *his* solutions. If we are lucky, the problem eventually goes away. If not, problems keep going round the organization without ever being fundamentally and permanently solved.

> In a large, sales-driven company, the problem of collating the information throughout the organization and using it for insight into the markets has survived almost ten years of trading. The company has a bounty of resources and systems, and information is ever-present in one form or another. However, there is no real market research and analysis at the service of line managers. For some reason, the line managers are not interested in these analyses, preferring to rely on the reports of their own sales forces. Consequently, the 'market research' department has been restructured a number of times (about once every two years) without any improved results. Once in a while the question arises 'should we axe the whole department?' A senior manager then decides to try to do something about it, restructures the department again, and six months down the line finds himself back where he started. The structural reasons that lead line managers to dismiss such information efforts, however, remain untouched because 'you don't change a winning team'.

We need to change the *thinking* that created the problem in the first place

Problem solving is not enough, because even if one is lucky enough to solve the problem permanently, the thinking that created the problem is still active, and still wrong. The real solution lies in the way we design the structures and operational policies of our organizations. To quote Jay Forrester, the founding father of the field of system dynamics: 'The role of senior management, especially of chief executive officers, should be that of corporate designers and not corporate operators.'[11] As designers, we must be able to have an understanding of how each part of the organization affects any other part. Outcomes result from the whole system, not just from each of the parts. The results of the parts can be wonderful and the total outcome a disaster. Our policies must take into account the effects on the organization *as a whole*, rather than simply an ailing operating part.

A need for right-brain thinking

We might conclude that there is a definite need for a right-brain capacity in left-brain cultures. As Charles Hampden-Turner points out, the Western preference for analytical thinking and individualism has many advantages in the early stages of industrial development;[12] analysis is a reliable way to understand simple mechanisms: by dividing things up, it shows their workings when their function is essentially a sum of parts put to a purpose. Yet, as organizations grow to global market sizes, as products become a complex sum of benefits to widely different customers, analysis is then of less value. When the focus shifts from machines to social systems, there is little to gain from dividing the whole into parts. Our so-called rationality is no longer up to the challenges of the world that we ourselves have created, with its integration, complexity and turbulence.

In the field of management, several researchers have recognized the limits of the classical left-brain approach. H. Mintzberg, in an article titled 'Planning on the Left Side and Managing on the Right', argues that whereas most planning is done with a distinct left-brain approach, research on senior management processes suggests that they possess characteristics of right-brain thinking:

> First, I would not like to suggest that planners and management scientists pack up their bag of techniques and leave organizations, or that they take up basket-weaving or meditation in their spare time. (I haven't—at least, not yet!) It seems to me that the left hemisphere is alive and well; the analytic community is firmly established, and indispensable, at the operating and middle levels of most organizations. Its real problems occur at senior levels. Here analysis must coexist

with—perhaps even take its lead from—intuition, a fact that many analysts and planners have been slow to accept. To my mind, organizational effectiveness does not lie in that narrow-minded concept called 'rationality'; it lies in a blend of clear-headed logic and powerful intuition.[13]

Yet, right-brain thinking is not an answer

The main problem is that the alternative, right-brain thinking, is not a complete solution. As we have seen, right-brain thinking is mostly non-verbal; it perceives and processes images and feelings and is difficult to articulate and therefore to communicate to others. Concurrently, the synthetic and holistic capability of right-brain thinking is necessary, but difficult to put into practice. How can organizational systems deal with the interconnectedness of all things? Furthermore, right-brain approaches are intuitive, timeless and diffuse. As such, they seem very distant from everyday business problems and concerns. One does not see clearly how they can be integrated into more traditional left-brain thinking. The Japanese example is striking, but these elements have been deeply embedded in Japanese culture through religion, philosophy and social practices for centuries. How can we reject our cultural heritage to make room for more right-brain attitudes? Should we convert executives to Zen Buddhism?

We need to make rational decisions

In organizations, the need for rationality is overpowering. The main reason is still that, although limited, the 'rational' process in decision making is better than any other we know. Analysis is necessary in business, at all stages and in all functions. Particularly when the stakes are high and they involve a number of other people, we want to be certain we are making the best possible decision at the time.

This need is reinforced by the accountability structure of organizations. Unless we are at the very top of our own company, in most positions we are accountable to somebody for our actions, whether to our boss, to the board, or to our own staff. This accountability implies that, firstly, the decision maker needs to have some sort of understanding of the implications of his or her decisions, and secondly, that these decisions can be justified in a way that will generally be accepted. People might disagree with choices, with weightings given to various factors, but they will not throw the book at us because we made a 'completely irrational choice'. This need for justifying one's decision is a strong motivator for a robust decision-making process.

Furthermore, decisions in organizations generally involve more than one person. For the stakeholders to 'buy into' the proposed choice, they need to understand it. A rational process is a convenient way to communicate the ins and outs of a decision. Again, people will argue, but they will speak a common language we might call 'rationality'. Confronted with an intuitive or emotional decision, stakeholders will feel excluded, because intuition, for instance, is a typical right-brain process, it is difficult to verbalize, let alone to explain.

A strong 'rational' culture will be difficult to change because of this need to show results, accountability and communication. All other aspects of corporate culture will tend to maintain one set pattern, particularly when it is so deeply ingrained that it is accepted as reality, rather than as one of many possibilities. Unfortunately, as we have seen, what is presently understood as being 'rational' conveys its own problems. In some cases the negative side-effects of this left-brain approach far outweigh the positive benefits expected from the chosen course of action. And yet, we are caught in a double bind. Although the need for right-brain thinking is obvious, it seems so opposed to anything we do, say or think, that it is difficult to see how we are going to integrate it into existing structures.

Overview

In this chapter we demonstrate the value and limits of 'classical' rationality in companies. We attempt to show that a purely financial outlook on things can turn a guardian into a gaoler. Although financial approach may appear 'rational', it can be a poor guide for decision making. We then explore the need for more 'right-brain' thinking in business, while maintaining a down-to-earth analytical approach to business situations.

Notes

1.   See A. Hirschman, *The Passions and the Interests*, 1977, Princeton University Press.
2.   D. Bohm and M. Edwards, *Changing Consciousness*, 1991, HarperCollins.

3. For more on decision analysis see: H. Raiffa, *Decision Analysis*, 1961, Addison-Wesley.

4. See J. March, *Decisions and Organisations*, 1988, Blackwell.

5. R. W. Sperry, 'The great cerebral commisure', *Scientific American*, January 1964.

6. F. Trompenaars, The organization of meaning and the meaning of organization. Thesis, the Wharton School of the University of Pennsylvania.

7. C. Hampden-Turner, *Corporate Culture: From Vicious to Virtuous Circles*, 1990, Random House.

8. For a very shrewd analysis of that tragedy see: C. Argyris, *Overcoming Organisational Defenses*, 1990, Simon & Schuster.

9. R. Pascale, *Managing on the Edge*, 1990, Simon & Schuster.

10. J. March and J. Olsen, 'Garbage can models of decision-making in organizations' in J. March and R. Weissinger-Baylon (eds), *Ambiguity and Command: Organizational Perspectives on Military Decision-making*, Ballinger.

11. In M. Keough and A. Doman, 'The CEO as organisation designer', *The McKinsey Quarterly*, 1992, No. 2.

12. C. Hampden-Turner, *Corporate Culture: From Vicious to Virtuous Circles*, 1990, Random House.

13. H. Mintzberg, *Mintzberg on Management*, 1989, Macmillan.

Chapter 2

Mental models

Although thinking is generally accepted as being a difficult and challenging activity, we all feel that we think very well. In fact thinking seems so natural that we take it completely for granted. Yet, thinking is not so naive and innocent as it seems to be. Although we are often capable of great creative leaps, most of the time our thinking is pretty 'conditioned', our ideas follow well defined tracks, inroads in our minds that we have built there through experience, knowledge or simple belief. The architecture of our thinking is taken for granted even more than thinking itself. By architecture, I mean the very way we order our thoughts, construct our mental representations of reality around us, in other words, the way we build our 'mental models'.[1] Mental models are deeply held, often subconscious sets of assumptions about how the world works. They affect our perception and evaluation of the situations we encounter.

> They met in a bar where he offered her a ride home. He took her down unfamiliar streets. He said it was a short cut. He got her home so fast she caught the ten o'clock news.[2] Why is the ending to this story so surprising? Why do we expect that this woman is getting into trouble? The beginning of the story triggers an unspoken assumption that makes us jump to a 'trouble' conclusion. In fact, it is more than an assumption, what is acting here is what we call a 'mental model'. (In this instance a 'don't talk to strangers' mental model.)

These 'theories' or 'images' are what we will call mental models—or mindsets, theory-in-use and so forth. Such mental models tend to follow three general rules: consistency, stability and simplification.

A mental model is internally consistent

Most people find it difficult to maintain contrary beliefs. In situations when they have to do so the result can be intense stress and anxiety created by a 'cognitive dissonance'.[3] This often happens when a person's experience is contrary to his or her beliefs.

Mental models are stable, and tend to resist change

People tend to hold on to their beliefs, attitudes, opinions and even prejudices. When something happens to contradict their mental model (cognitive dissonance) the first reaction is to dismiss what happened so as not to modify one's mindset. If it cannot be dismissed then we may finally modify our convictions.

> Some attitudes are remarkably stable over time. One of the most persistent attitudes of the workplace is that managers *think* and workers *do*. Obviously, in many instances the person on the job will know more, and have better ideas about how it works than his or her manager. Yet, managers are implicitly convinced that their seniority confers on them greater knowledge. Although this attitude has been denounced and is constantly shown as reactionary, its longevity is surprising. It is so deeply set in our mental models that most of us find ourselves telling subordinates what to do—even in areas where they have proven that they are more competent than we are!

Mental models are simplifications of the real world

They work as 'maps' of reality, and as such are often over-simplified, or do not focus on the relevant aspects of reality. Yet, they are necessary for us to deal with the complexity of the world. The danger of this simplification is the confusion that arises between the model and the thing itself. The map is not the territory, the word is not the thing (*the word cat does not scratch or purr*). Our right brain, however, tends to confuse the map and the territory. Philosopher Gregory Bateson[4] uses the following example:

> If a patriotic person meets someone engaged in the action of burning his national flag, he or she is likely to walk up to the flag-burner and punch him on the nose. Why would anyone want to hit somebody for burning a piece of cloth? The answer is that the right brain doesn't make any difference between the flag and the emotional values it embodies. In a way, for our right brain, burning the flag is burning the country (after all, why would the other person be burning the flag in the first place?)

This confusion is part of our emotional make-up but can be extremely misleading at times and lean towards dogma rather than empiricism. If we are emotionally attached to an idea, such as religion, free will, or saving the dolphins, we will have great trouble facing opposing possibilities such as 'Could God not exist?', 'Is Fascism a worthwhile political system?' or 'Are dolphins actually safe from extinction?' For the left side, understanding does not mean agreeing but for our right side, the emotional content of the 'if' is the same as the action itself. After all when you have to go for a vaccine shot, is not the pain at the thought of the needle worse than the injection itself?

Mental models are stories that run through our minds

Mental models are unstructured little scenarios, or scripts, that run through our minds. They represent our view of the world, and of the forces that act on it. In fact, they are quite welcome, because with the universe being the weird and complex place it is, we would be very lost without them. They are the basic units our minds use to store and manipulate information (an upgraded version of the 0 and 1 of a computer). Mental models are efficient because they can be very loose and unstructured. They can appear either as images or sentences, or sometimes even as feelings. These models run through our minds at all times creating the structure of our 'internal dialogue'. A mental model is the result of subconscious structuring of observed reality into a manageable form.

> Gregory Bateson[5] used to tell the story of the man who had built the Ultimate Computer. One day, this man asked the machine to compute whether one day it would think like a human being. The computer computed for days and nights till it finally flashed the following answer:
>
> 'Ahhh ... that reminds me of a story ...'

Sometimes stories, sometimes less structured 'images', it would seem that our mental models are organized around four main elements of representation: background, experience, context and purpose.

Background

Background represents the sum of what we have learned 'abstractly'. It regroups our religious, ideological and political attitudes towards the world and how it works. Most of our 'opinions' can be traced back to our upbringing, particularly in abstract matters where direct experience is almost impossible.

Experience

By experience, I mean all our conditioning: both the values that we are brought up with and the memories of our own personal experience. This conditioning can be very strong, and hard to break. It comprises many attitudes and opinions that we have accepted from other people and that we cling to although direct observation seems to point otherwise.

Context

The environment at the time of the event. Profit of 5 per cent after tax this year will have an entirely different meaning if we have made 5 per cent each year for the last ten years or if we have been making losses for the five previous years. Similarly, we will interpret a friend's laughter differently depending on whether we are at a dinner party, at a meeting, or at a funeral.

Purpose

We are usually greatly influenced by our current concerns. For instance, a friend of mine just bought a VW Golf and keeps pointing out all the VW Golfs we pass in the street. In the same way a smoker will notice tobacconists whereas a non-smoker will not. On a corporate level, for instance, if you have opted for a growth or expansion strategy you will tend to pick up those indicators which show that recession is nearly over. As a rule, chief executive officers (CEOs) tend to understand their economic context in favour of their current strategies.

If the only tool you have is a hammer, you tend to treat everything as a nail. Let us try the following exercise.

FINISHED FILES ARE THE RESULT OF YEARS OF SCIENTIFIC STUDY COMBINED WITH THE EXPERIENCE OF YEARS.

Now, count the times the letter F appears in that sentence. Count them only once, do not go back and count them again. Record your answer here:_____

Now state your confidence in your answer on a scale from 0 per cent meaning that you are sure that you are incorrect, to 100 per cent meaning that you are sure that you are correct. Record your confidence here: _____

Well, there are six F letters in the sentence. Most people get it wrong (I only saw three Fs when I took the test), but that, in itself is not a great problem. The point is the confidence people have in their answers. American researcher J. Scott Armstrong[6] has been conducting this test with a sample of 50 and reports an average confidence level of 91 per cent. For the 34 per cent who had the correct answer, the average confidence level was about 87 per cent. For the 66 per cent who had incorrect answers, the average confidence level was about 93 per cent. Imagine how confident people are going to feel about something they have been doing all their lives, regardless of the fact that it might not work any more. Because we are so confident in our thinking, we need to make mental models explicit to be able to share and discuss them with other people involved in our decisions.

Mental models determine how we act

By framing 'how we see the world', our mental models determine not only how we make sense of the world, but how we take action. Harvard's Chris Argyris, who has worked with mental models and organizational learning for 30 years, puts it this way:

> Human beings have programs in their heads about how to be in control, especially when they face embarrassment or threat, two conditions that could lead them to get out of control. These programs exist in the human mind in two very different ways.
>
> The first way is a set of beliefs and values people hold about how to manage their lives. The second way is the actual rules they use to manage their beliefs. We call the first, their espoused theories of action; the second their theories-in-use.[7]

Years of research show that although most people do not always behave consistently with their espoused theories (what they say), they do behave according to their theories in use (their mental models).[8] The distinction is vital to understand mental models. Often we tend to take at face value what people tell us about how they see a situation. As a rule, it is safer to observe how they act about situations to complete our understanding of their mental model. Mental models are active—they shape how we act. They affect what we see and how we think we should react. The problems with mental models lie not in whether they are right or wrong—by definition all models are simplifications. The problems with mental models arise when models are tacit—when they exist below the level of awareness.

> If you ask board members or top executives what their main role is in the organization, they tend to answer that their prime role is one of strategic management. They are quite explicit that they should spend most of their time together, exploring and resolving strategic issues. However, each time they meet, they spend 80 to 90 per cent of their time performing symbolic and legitimating functions, handling administrative procedures or reviewing what has already happened. Consequently, strategic issues tend to be shelved or postponed till the next meeting. When confronted with such a statement of 'fact', these top managers tend to deny or dismiss it: it does not fit with their mental model of themselves as top executives.

There is nothing wrong with mental models themselves—it is simply the way our mind works. The fact is that we could not function without them. Where computers use zeros and ones to think, we use images, mental models, theories in use, microcosms: complex sets of ideas that are triggered by circumstances. These structures are necessary 'short cuts' in thinking, unless they become fixed and/or hidden and denied. Confusing mental models and reality is similar to confusing map and territory. Maps are notoriously simplified and unreliable. The territory is something else altogether. We need the map to deal with the territory, but mental sanity depends on our keeping map and territory separate.

Thinking follows the principle of least effort

Our use of mental models tends to follow the principle of the *least mental effort*. When the mind is confronted with complexity, it tends to fall back on information-processing tricks to reduce cognitive effort. Our three main 'short cuts' are given below.

The use of stereotypes

Norwegians are tall and blond, cars pollute, children are noisy, exercise is healthy, and so forth. Stereotypes are frequently abusive generalizations that we use as 'propositions' in our reasoning. Most stereotypes can be proven wrong, and yet most stereotypes are by and large true. (Does the presence of a black swan invalidate the idea that swans are white?) However, stereotypes are dangerous when totally unrelated to experience. My stereotypes about Eskimos are likely to be very wrong since I have never met an Eskimo. The danger is to apply stereotypes in the face of evidence to the contrary. In the business world some stereotypes have gained a remarkable longevity—and caused no end of grief. Consider, for instance the following two assertions:

> Hardly a competent workman can be found who does not devote a considerable amount of time to studying just how slowly he can work and still convince his employer that he is going at a good pace. Under our system a worker is told just what he is to do and how he is to do it. Any improvement he makes upon the orders given to him is fatal to his success.
>
> (Frederick Taylor)[9]

> We are going to win and the industrial West is going to lose out; there's not much you can do about it because the reasons for your failure are within yourselves. Your firms are built on the Taylor model. Even worse, so are your heads. With your bosses doing the thinking while workers wield the screw-drivers, you're convinced deep down that this is the right way to run a business. For you the essence of management is getting the ideas out of the heads of the bosses and into the hands of labour.
>
> (Konosuke Matsushita)[10]

Both are evidently stereotypical, and both, one can argue, as crazy in their extremism as each other. Yet, the thinking behind these two quotes has structured the larger part of business practices since the 1920s and again in the 1960s. Such stereotypes run 'deep and true', but remain nevertheless stereotypes. Unless we learn to free ourselves from such outrageous convictions, there is very little chance of achieving rational thought.

Thinking via analogy or metaphor

The business world is riddled with anecdotes, short stories, parables and the like. When a situation is too complex, we try to represent it by a more commonplace, simple occurrence. Metaphors can be very powerful if they 'sound good'—and equally can be misleading. We use an analogy or metaphor when we say 'it is like ...' Here are some of the most common parables associated with systems thinking:

> *The boiled frog*: If you place a frog in water, and slowly increase the temperature, the frog will not notice the difference and let itself be boiled to death. Industries or large companies are often compared to 'boiled frogs'.

The blind men and the elephant: Four blind men who do not know what an elephant is, are led close to one of the beasts. The first blind man grabs the trunk of the elephant, the second his leg, the third his side and the fourth his tail. When asked to describe what an elephant is, the first blind man describes it as a large snake, the second as a tree, the third as a wall and the fourth as a brush. This analogy is often used to express the feeling that unless one perceives the system as a whole, one might have a very biased vision of what the parts do.

The frog and the scorpion: The scorpion wants to cross the river, but cannot swim. It asks the frog if it would be willing to carry it across on its back. The frog answers in no uncertain terms: 'are you crazy? if you climb on my back you will sting me and I will die!' 'No', argues the scorpion, 'why would I do such a thing. If I sting you and you sink, I'll drown with you.' Swayed by the logic of this argument, the frog accepts and carries the scorpion into the water. In the middle of the stream the scorpion stings the frog. 'Why?' asks the dying creature, 'you shall now die with me.' 'Alas' answers the drowning scorpion, 'it's in my nature to sting.' This story exemplifies that systems do what they are supposed to do, not what their leaders want them to do.

The mullah and the key: A man saw Nasrudin searching for something on the ground in the middle of the street 'What have you lost, Nasrudin?' he asked. 'My key' says the Mullah. So the man goes down on his knees and helps him look for it. After a time, the other man asks 'Where exactly did you drop it?' 'In my house' answers Nasrudin. 'Then why are you looking here?' 'There is more light than inside my house.' Draw your own conclusions![11]

The use of rules of thumb

Rules of thumb are a more elaborate type of short cut, because they include an if ... then. Where stereotypes, metaphors and analogies tend to be general statements (*all* swans are white), rules of thumb tend to be conditional. They are usually built around 'critical indicators' that should describe a situation.

> A production process improvement consultant visits a plant with some 'indicators' in mind such as: How many people are not doing anything at one given time? What proportion of those who are doing something are doing value-added work? How clean is the plant? How many forklifters are around? How well is the space utilized and so forth. The answers to these questions give him rules of thumb to know what he is going to tackle first, and how.

Each of these short cuts represents a 'point of view' on the situation. Often, thinking something through is simply too costly or too difficult. However, we must be aware that we use these simple rules even when they are not appropriate to what is actually happening out there!

The key to effective thinking is to use several of these short cuts on the same problem to see if a solution emerges that is consistent with these different points of view.

Mental models need to be made explicit

Actually, getting mindsets or mental models to the surface of our thinking is rather difficult. Expressing our hidden assumptions can even be frightening. Chris Argyris argues that we trap ourselves in 'defensive routines' that insulate our mindsets from examination. In consequence, we develop 'skilled incompetence'—what Argyris describes to be the attribute of people who are 'highly skilful at protecting themselves from pain and threat posed by learning situations'.[12]

> Defensive routines are so diverse and commonplace that they usually go unnoticed.

> We say, 'That's a very interesting idea', when we have no intention of taking the idea seriously.

> We deliberately confront someone to squash an idea, to avoid having to consider it.

> In the guise of being helpful, we shelter someone from criticism, but also shelter ourselves from engaging difficult issues.

> When a difficult issue comes up, we change the subject.

The classical mental model

A classical thinker will tend to build his mental models along certain principles of 'classical rationality'. These, as we have seen, revolve around the idea of looking at each element of a problem independently from the others, and at one given point in time.

Challenging assumptions is difficult, but increasingly held to be good practice. Where *systems thinking* differs from most 'thinking' processes, is that once the assumptions are made explicit, it sheds a new light on the causality between assumptions and actions in our mental models. Ever since Descartes, we have assumed that the best method for looking at a problem is to break it down into its elementary components. The very logic of classical rationality shows strong in-built assumptions that will affect the outcome of our thinking. We assume that the whole is nothing more than the sum of the parts, and therefore to understand it we need to understand each of the parts separately. This is the original meaning of *analysis.*

We tend to ascribe an explanation of something to specific, out-of-the-ordinary events. A stock market crash can be traced to what someone said; as can last month's sales, new budget cuts, who just got promoted or fired, etc. This can lead us into 'external' explanations. Something 'out there' is responsible for what happens to us.

The classical thinker tends to use linear causality: this is a flow-chart mentality; we will try to draw chains of events and consequences under the assumption that a cause can be distinguished from its effects. In his introduction to the Ithink software,[13] Barry Richmond asks the following question:

What causes companies to flounder?

Think about it for a few seconds and please provide an answer in the space below:

Laundry lists

When faced with a problem, our natural reaction is to make a 'laundry list'. Your answer is in fact your mental model of what causes companies to flounder. I have asked a variety of people the same question and most of the time they come up with a list of factors such as:

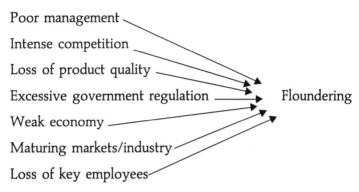

This is what Barry Richmond calls 'laundry list logic'. Yet, whereas poor management can certainly cause floundering, it is very likely that floundering will also create poor management as managers panic and try to cut corners to save the day (Figure 2.1).

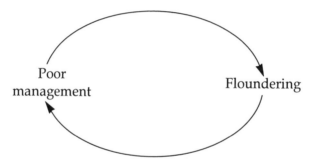

Figure 2.1 Poor management leads to floundering, which reinforces poor management

Laundry list logic will lead to laundry list solutions
Similarly, with the loss of key employees or loss of product quality, if we try to represent these relationships we end up with a very different picture (Figure 2.2).

With 'laundry list' logic, we would have tried to sort out management, or product quality, or tried to retain key employees, whichever we felt was the critical factor. This stems from a static view of the world. In a dynamic view of the world, causality is no longer one way. Cause creates effect that becomes cause, and so on.

Shift of mindset

These classical mental models are clearly out of sync with the complexity of our present business environments: the world has

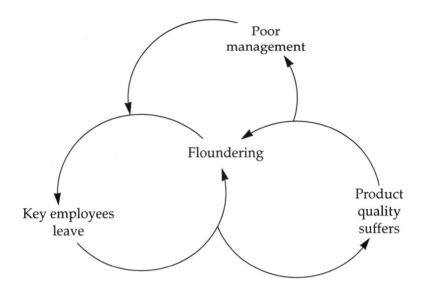

Figure 2.2 The dynamics of failure

changed—but not our assumptions. Therefore we need to realign our thinking with 'reality'.

Think global!
A company cannot remain 'local' and compete in the 1990s. The entire business environment is more integrated. We need to work with our suppliers, with the regulatory bodies and consumer pressure groups. To build competitive corporate cultures in our current context, managers must pay close attention to all their constituencies—such as customers, shareholders and employees. Business is increasingly interdependent, we cannot act in isolation any more. To quote Barry Richmond:

> The action is between, not within! It's between functions, between suppliers and producers; between competitors; between organizations and the community, and between companies and the eco-system. In the 90s, the river no longer purifies itself every two miles. To wit, there are few two miles stretch left because today, everyone lives downstream![14]

Organizational behaviour is dynamic, not static. We have to learn that stability is an illusion—and not even desirable. Organizations must live in a fast moving environment and their only hope of survival is to be adaptable. To deal with this sort of change, we must learn to think dynamically rather than statically.

Discontinuous actions
The underlying assumption is that one can take discontinuous action and hope that all the other parameters will 'sort themselves out'. This belief is at the root of many consultancy assignments where consultants come in, fix the problem and get out, while the client company does not fare any better.

Internal or external focus
Another consequence of laundry list logic is an implicit focus on external explanations. External forces usually seem so much stronger and easier to blame than internal forces that they tend to bear the blame. However, experience shows that most of the problems companies have can almost always be traced to the way they *do* things—not to the customers, to the competitors, to the regulators or to any other convenient scapegoat.

Overview

In this chapter we have outlined a simple approach of how our minds work: by storing information under the form of mental models. These mental models are pictures or stories that represent our assumptions about how the world works. They tend to be stable, consistent and over-simplified. Our mental models are built upon our education, experience, purpose and context, and we use them in specific ways. We need mental models to think, but unless we are aware of them—and how they influence our thoughts and actions—we remain their prisoners. Therefore, our aim is to manage our own mental models.

Notes

1. For a general presentation of mental models, see P. Senge, *The Fifth Discipline: The Art and Practice of the Learning Organisation*, 1990, Bantam Doubleday. For a more technical discussion of mental models see P. Johnson-Laird, *Mental Models*, 1983, Harvard University Press.
2. See R. Fisher and W. Ury, *Getting to Yes*, 1981, Hutchinson.
3. L. Festinger's theory of cognitive dissonance (L. Festinger, *A Theory of Cognitive Dissonance*, 1957, Row, Peterson) is founded on systems thinking, and has been represented in terms of feedback loops by G. Richardson (G. Richardson, *Feedback Thought in Social Sciences and Systems Theory*, 1991, University of Pennsylvania Press).
4. G. Bateson, *Mind and Nature*, 1979, Wildwood House.
5. Ibid.

6. J. Armstrong, *Long Range Forecasting: From Crystal Ball to Computer*, 1985, John Wiley & Sons.

7. C. Argyris, *Overcoming Organisational Defenses*, 1990, Simon & Schuster.

8. C. Argyris and D. Schon, *Theory in Practice*, 1974, Jossey-Bass.

9. In R. Pascale, *Managing on the Edge*, 1990, Simon & Schuster.

10. Ibid.

11. From I. Shah, *Wisdom of Idiots*, 1971, E. P. Dutton.

12. C. Argyris, *Strategy, Change and Defensive Routines*, 1985, Pitman.

13. Barry Richmond, *Ithink User Guide*, 1990, High Performance Systems.

14. Ibid.

Chapter 3

An alternative rationality: systems thinking

The systemic approach

However, if we consider this contradiction more closely we will find that we take for granted that the 'rational' approach will be the one we presently use: the 'financial decision'. In effect, this form of thinking is extremely left brain: it is reductive into parts; it insists on measurable quantities, it has a static vision of the world; and it is predominantly sequential. Yet, is it rational? In the true sense of the term, something rational is 'of the reason', endowed with reason. Reason is not necessarily left brain. Actually, many of the attributes of left-brain thinking are definitely unreasonable—expecting that the whole is nothing more than the sum of the parts, believing that we can change one factor without affecting others, thinking that anything real must be measurable, etc. Our entire experience of life (as human beings) tells us that this is not true. Yet, to be reasonable something must be verbalized, analytic, realistic, and so forth. Reason does not have to be limited to left-brain thinking. Reason has to be logical. Logic itself is nothing more than consistency with explicit assumptions. It is therefore perfectly feasible to build a logical framework that will not contain the inherent limitations of the traditional left-brain approach.

To understand things, we take them apart and study the pieces. To improve things, we try to improve each of the pieces individually. It is rather like trying to get a horse to run faster by teaching each of its legs to perform a more efficient movement. The legs might learn how to work perfectly, but the moment the horse is back on the grass and tries to gallop, it is not likely to go very far. The systems approach focuses on the interrelationships, how the horse's legs relate to each other and back to the horse. An organization remains an organization even when every person and machine has been changed, as long as the purpose, hierarchies and systems of reward and punishment are the same.

This is the basis of systemic reason as opposed to traditional reason. It is perfectly rational, it just includes more elements than traditional thinking in so far as it remains verbal and analytic as well as synthetic, holistic and dynamic. The systems approach provides a superior rationality to our traditional ones. It accepts non-measurable elements, in its very nature it deals with dynamic behaviours and its focus is not on the individual parts but on the interrelationships between these parts.

> This approach is also inherently Western. Several authors have traced systems concepts through the history of Western Science. Mayr, for instance, traces the history of feedback to Ancient Greece; he credits Ktesibios (250 BC) with the first known feedback device.[1] Although the term 'system' itself was not emphasized, the history of the concept can be traced back to Leibnitz; to Nicolas of Cusa with his coincidence of opposites; to the alchemy of Paracelsus; to Vico's and ibn-Khaldun's vision of history as a sequence of cultural entities or 'systems'; to the dialectic of Marx and Hegel, to mention but a few names from a rich history of thinkers. The systems approach is already part of our culture and as such, we can all relate easily to it.

As such, systems thinking is a conceptual framework. Rather than viewing an organization as driven by a set of 'factors', most of them external, we represent it as the continuous interplay of the interactions between elements. Because the behaviour of a system is largely generated by those interactions between its elements, the people within the organization are therefore capable of exerting a significant influence on it by modifying some of the relationships.

Origins

Systems thinking draws from more than a century of theory and practice. The origins of the field can be traced back to engineering theories about feedback control first published in 1867 in 'On governors', a paper by the Scottish physicist James Clark Maxwell.[2]

> In the nineteenth century, engineers had managed to adapt some sort of governor to the steam engine but still ran into some problems. They came to Clark Maxwell with the complaint that they could not draw a blueprint for a self-regulating engine. Their problem was that they had no theoretical understanding of the problem that could help them predict how the machine would behave once built. There were several sorts of behaviour: machines would go into runaway, exponentially increasing their speed until they broke down or slowing down until they stopped. Others oscillated and seemed unable to settle. Others still would seem to behave totally randomly.

Maxwell examined the problem and wrote out formal equations for relations between the variables at each successive step in the sequence. He hit the same snag that held up the engineers: combining this set of equations would not solve the problem. Finally he realized that the engineers were not considering time. As the ancient paradox of Epimenides:

'EPIMENIDES IS A CRETAN
WHO SAYS CRETANS ALWAYS LIE'

is a paradox only if you consider as a whole—if you involve time, statement by statement your answer would be yes … no … yes … no and so forth—the equations made no sense as a whole. However, by including time into the problem one could develop a workable answer to the problem—which led to the creation of the field of Cybernetics and Systems Dynamics.

The theory developed slowly until the 1930s, when Bell Laboratories developed electronics for the amplification of signals that depended on the stability of feedback systems. At the same period, a group of biologists at Vienna University were applying the same concepts to understand the growth of populations under various limiting conditions, and more generally the behaviour of 'ecosystems'. Out of these two parallel movements emerged cybernetics on the one hand through the pioneering work of Norbert Wiener at the end of the Second World War, and on the other hand, the mathematical basis for a 'general system theory' under the direction of Ludwig von Bertalanffy in the 1950s. His publication of *General Systems Theory* in 1968 helped to spread the basic concepts and ideas.[3]

From the engineering field, after working on control systems for stabilizing radar antennae on ships and the control of guns during the Second World War, Jay Forrester joined the MIT to be involved with the development of early digital computers. In 1956 he joined the recently created School of Industrial Management at MIT (later renamed the Alfred P. Sloan School of Management) to direct a systems dynamics program that explored the causes of industrial dynamics,[4] such as the reasons for success and failure of businesses.

He constructed the first computer-based simulations of industrial dynamics which opened the field for numerous attempts, with varying degrees of success, at computer modelling of social systems. From his work with various corporations, he discovered by modelling the structure of their decision processes that most problems (falling market share, unstable inventory, inadequate quality control) were traceable to the way the corporation did things rather than external circumstances. The MIT group has since developed models and methodologies that make systems thinking a practical, rational approach to analysis.

Analytical

Due to its origins in both biology and cybernetics, the systems approach is highly analytical. Although from a practical point of view many of its conclusions can be used as a thinking framework, most of the work which has been done in this field has been built on the basis of extensive mathematical representations. After the Second World War, it was gradually realized in the scientific world that systems theory was an attempt at scientific interpretation and theory where previously there had been none. General systems theory responded to a trend in various disciplines, and its applications multiplied in many diverse fields. In the late 1940s the development of computer technology, information theory and cybernetics was built on the premises of the systemic approach. Three fundamental contributions appeared at about the same time: Wiener's cybernetics,[5] Shannon and Weaver's information theory[6] and von Neumann and Morgenstern's game theory.[7] Wiener carried the cybernetic, feedback and information concepts far beyond the field of technology and generalized them in the biological and social realms. The systems framework has been consistently used in all those fields, albeit in a static way. While cybernetics had a predominant role to play in the development of technology, the systemic, or 'structural' approach has been widely used in psychology and sociology—hence our common understanding of 'The System'.

Classical system theory applies classical mathematics, i.e. calculus. Its aim is to state the principles which apply to systems in general, to provide techniques for their investigation and a vocabulary for their description. Due to the generality of such description, one might say that certain formal properties will apply to any system, even when its particular nature, parts, relations, etc., are unknown or not investigated. Examples include generalized principles of kinetics applicable to populations of molecules and biological entities, i.e. to chemical and ecological systems; diffusion, such as diffusion equations in physical chemistry and in the spread of rumours; applications of steady state and statistical mechanics to traffic flow;

allometric analysis of biological and social systems. Furthermore, to complete the development of classical systems theory, much work has been done with computerized simulations. Sets of simultaneous differential equations as a way to model or define a system are, if linear, tragically tiresome to solve even in the case of a few variables. If non-linear, they are simply unsolvable. For this reason, computers have contributed to the development of systems theory by facilitating calculations and by opening up fields where no mathematical solutions exist.

The 1980s have permitted the expansion of these techniques in the business world with the PC revolution and with simulation packages that run on portables. The greatest value contributed by computer modelling is to simulate how people view their own organizations. Because of the size and complexity of modern organizations, and the difficulties the human mind has to hold more than three or four variables at the same time, let alone to see dynamic interactions, one single person's representation is very likely to be wrong. We believe that the members of an organization will collectively have an accurate representation of the whole system, but individually they will only see the part with which they are most immediately concerned. By listing and diagramming the cause and effect relationships perceived by a range of people in the organization, and then using the computer to simulate dynamic behaviours, experience shows that we can represent rather accurately how the organization behaves on specific points. From then on, the decision makers can experiment with various policies to see which will have an effect, and which will not, and as a result can make better informed decisions about their companies.

Holistic

As such, 'things' do not exist in isolation. People, organizations, societies, problems, even jokes can only exist as part of a context. In fact, our mind simply does not understand things 'out of context', it has to relate facts or ideas with a general context to be able to comprehend them. However, most left-brain analyses consider things in isolation. This is one of the core elements of the traditional scientific approach: to take whatever we study in a nice, clean laboratory, in a controlled environment, and then cut it into pieces to see what it is made out of. Although Heisenberg's uncertainty principle states that the mere action of observing a phenomenon will influence this phenomenon, we still tend to isolate things to study or think about them. Unfortunately, most things lose any form of sense when taken out of their context.

> As Bateson notes,[8] consider a blind man walking with a stick. Where does the blind man's self begin? At the tip of the stick? At the handle of the stick? And what about when the blind man is eating, his stick resting next to him on the table? The answers to these fairly simple questions are bound to be confused if we consider the man and the stick in isolation. From a systems perspective, the stick is a pathway along which information about the street is transmitted to the man. To draw a line across this pathway is to cut off a part of the circuit which determines the blind man's movements.

To quote Donella Meadows, one of the pioneers of systems thinking and systems dynamics modelling in the United States:

> The bottom line message of the global models is quite simple: the world is a complex, interconnected, finite, ecological–social–psychological–economic system. We treat it as if it were not, as if it were divisible, separable, simple and infinite. Our persistent, intractable global problems arise directly from this mismatch. No one wants to generate hunger, poverty, pollution, or the elimination of species. Very few people favour arms races or terrorism or alcoholism or inflation. Yet those results are consistently produced by the system-as-a-whole, despite many policies and much effort directed against them. Many social policies work: they solve problems permanently, but some problems consistently resist solution in many cultures and over long periods of time. Those are the problems for which a new way of looking is required.[9]

Too many solutions provided by a classical left-brain approach assume a fixed pie. The only way to win is by increasing our share, which means someone else's share will have to shrink. This fixed pie approach is very limited in terms of problem solving or decision making. To see not only things but also relationships opens our vision considerably. We can start looking for solutions in unexpected areas, instead of always assuming that the solution is going to be close to the symptom. Systems thinking is not 'holistic' in the sense that we should try to connect the crab to the ozone layer, to the Canadian mongoose, capital life-cycle and all these back to ourselves, but in the understanding that what is causing our immediate problem might be 'outside the frame', and if the cause is outside the frame, so is the solution.

Pragmatic

Systems thinking, by its very nature, is pragmatic. To understand how a system behaves one has to look at the operational reality, not the aggregated numbers. The 'holistic' approach does not mean a high view, dealing with averages and indicators disconnected from the operational reality of things, but that we include all relevant elements in our analysis. High level indicators are seldom any help to show where the problems actually lie: if we have a 10 per cent machine stoppage index in a workshop, does this mean that we have to replace all our machines because they are poor quality and down 10 per cent of the time, or are there two machines which keep breaking down constantly and only operate 50 per cent of the time? The aggregated indicator will not give us this vital information. Systems thinking will prompt us to go and look for ourselves at what is actually, operationally, happening.

> The most tragic effect of this measurement mentality can be seen in redundancy programmes. If someone decrees that the staff to turnover ratio is too high, the branches or departments of the organization will be asked to shed, say, 15 per cent of their staff. There is no doubt that the ratio does reveal inefficiency. Yet this inefficiency is going to be found in the working processes themselves. Most likely, 5 per cent of the work is really value-added work, whereas, the 95 per cent left is non-value added. Ironically, those accountable for the value-added work will be pressured for even more productivity, when the real productivity gains can be obtained from squeezing the non-value-added work out of the process. By doing so, the unit should 'naturally' shed staff, over time, permitting honourable treatments of traumatic events in any case. If we simply go ahead with a massive redundancy programme, the likely outcome will be that the department will stay as inefficient, with fewer staff to do the work, and the pressure will soon be on to hire new staff because those in place 'just cannot cope'.

Most systems are goal seeking in their nature; as long as the operational processes are not changed, regardless of whatever shock is given to the system, it will either get back to its original steady state over time, or else move in a 'chaotic', i.e. violent and unpredictable, manner.

In practice, systems thinking follows these guidelines:

- *Focus on the relationships rather than the parts*. A system is a set on interconnected parts. In our usual reductionist view of things the emphasis is on the parts. To understand things we take them apart and study the pieces. With systems thinking, the interrelationships are important. We can understand how elements of a system interact on each other to produce a global outcome. A powerful insight of systems thinking is in seeing how structure influences behaviour.

- **See patterns, not events**. Rather than accepting event-type explanations, we shall try to identify the ongoing patterns, in order to perceive the forces underlying these events. This will also help us to avoid the 'boiled frog' syndrome. Too often, if we only see events we might only identify what is happening when it is too late to do anything about it.
- **Use circular causality**. Causality is seldom one way. Cause becomes effect, which then becomes cause and so on. Whenever you postulate that A causes B, look for all the ways that B in turn affects A. Most causal chains will revert on themselves to create 'feedback'.

Overview

Understanding how mental models are built and operate enables us to free our thinking from hidden logical assumptions. Systems thinking challenges some of these assumptions, such as linear thinking, and focuses on the parts rather than the whole. Systems thinking proposes a logic based upon circular logic, focusing on relationships rather than on the elements themselves and seeing patterns rather than events. Originating from cybernetics and biology, systems thinking is highly analytical, holistic and pragmatic. In effect, it proposes an alternative rationality—to complete a more traditional vision of the world.

Notes

1. A float valve! See Otto Mayr, in G. Richardson, *Feed-back Thought in Social Sciences and Systems Theory*, 1991, The University of Pennsylvania Press.
2. J. Maxwell, On governors, 1868, *Proceedings of the Royal Society of London*.
3. L. von Bertalanffy, *General Systems Theory, Foundations, Development, Applications*, 1968, George Braziller.
4. J. Forrester, *Industrial Dynamics*, 1961, MIT Press.
5. N. Wiener, *Cybernetics: Or Control and Communication in the Animal and the Machine*, 1948, MIT Press.

6. C. Shannon and W. Weaver, *The Mathematical Theory of Communication*, 1949, University of Illinois Press.
7. J. von Neumann and O. Morgenstern, *Theory of Games and Economic Behaviour*, 1953, Oxford.
8. G. Bateson, *Steps to an Ecology of Mind*, 1972, Ballantine Books.
9. D. Meadows, 'Whole earth models and systems', *The CoEvolution Quarterly*, 1980.

Chapter 4

The systems thinking framework

The systems thinking framework uses a number of conceptual tools to represent and describe reality according to this new set of assumptions. A system is any set of interconnected elements. For a system to function as such all the parts must be present and functioning; the cooling system in a car may consist of a radiator, fan, water pump, thermostat, cooling jacket and several hoses and clamps. Together they keep the engine from overheating, but separately they are useless. To do the job, *all the parts must be present and they must be connected in the proper way.*

If something is made up of a number of parts and it does not matter how those parts are arranged, then we are dealing with a *heap*, not a system. Moreover, a heap is not essentially changed by adding to its size, nor by taking some parts away. Adding more sand to the sand already on the ground just gives you a larger amount of sand, but adding one cow to the one you already have does not give you a larger cow. On the one hand we are dealing with a heap of unrelated particles that constitute a whole, and on the other, with a complex biological system. Whereas the heap can be changed and still remain a heap, the system—with all its complex interrelationships—cannot.

A survey of mergers and acquisitions during the 1980s shows that only 18 per cent of the companies who initiated a merger or an acquisition thought it was worth while and would gladly do it again. Although the failure rate is known to be high, many companies have taken that decision. Why? It looks very good on paper: we can double our production capacity, our total assets, our value on the stock market will triple, and so will our client list and market share, etc. When we think like this, we are treating companies as heaps, not systems. The idea is that if you add two companies together, you'll get a bigger, happier company. Firms, however, are systems, not heaps. They have their own control loops, cultures and people. By trying to add them together, one can end up in a mess. For instance, General Motors acquired a small IT company called EDS. GM is manufacturing driven, with deep hierarchy, low margins and total risk aversion whereas EDS was marketing driven, with shallow hierarchy, high margins and

opportunity oriented. The result was a failure because the
members of both companies simply could not work together.[1]

The concept of feedback

One of the major concepts of systems thinking is that of feedback;
rather than thinking, as we usually do, that effect have a cause, such
as 'if A, then B' (Figure 4.1).

Figure 4.1 Cause and effect: if A, then B

Systems thinking takes into account the fact that if A causes or
affects B, then in many ways, B is going to affect A. This circular
causality is what is called a feedback loop (Figure 4.2).

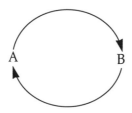

Figure 4.2 Feedback loop: A affects B, and B affects A

Representations derived from a systems thinking framework are
built on the interactions between parts rather than simple 'because'
causality. Instead of one-way causal statements, one is then induced
to build 'systems' or series of 'feedback loops' coupled with each
other. To take a simple example, consider a man felling a tree with
an axe. One-way logic will say: 'the man fells the tree'. The man uses
the axe, monitors the dent he is making in the tree and continues to
work until the tree falls. Each stroke of the axe is modified or
corrected, according to the shape of the cut face of the tree left by
the previous stroke. The man influences the tree (cuts it down) but

the tree also influences the man (until the tree is cut, he continues his hard work) (Figure 4.3).

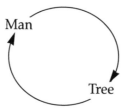

Figure 4.3 The man fells the tree, the tree controls the man

In fact, this is a self-corrective process involving a system, tree–eyes–brain–muscle–axe–stroke–tree; for obvious practical reasons, it is easier to say 'the man cuts down the tree' but it deludes us in thinking that the man carries out a linear action when, in fact, he is involved in a self-corrective system. There is a feedback loop from the tree to the man that enables him to continue cutting and know when to stop: the man's goal is to cut the tree, his action is to strike it with his axe, and the strength and number of strikes is determined by the gap between the size of the dent made in the tree and the tree being actually cut down (Figure 4.4).

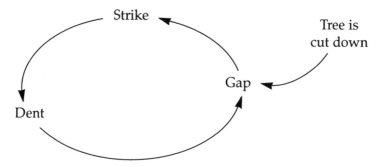

Figure 4.4 A goal-seeking system: until the tree is cut down, the man continues to strike

Linear causality

The feedback loop often goes against deeply ingrained ideas such as linear causality. The linear notion of cause and effect is deeply anchored in our thinking, and has been so ever since Greek philosophers started wondering about it. Since antiquity, and through all Christianity causality has been seen as a tree of causes that can be traced to one original cause—which has been given several names. Our very language is structured around the notion of some sort of linear causality (subject, verb, adverb). For instance, I will say:

'I fill the bath.'

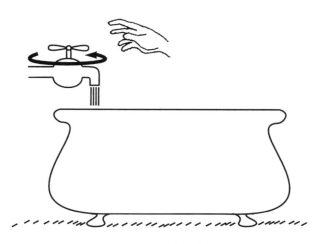

Figure 4.5 'I fill the bath'

In truth, the mental model conjured by this sentence is clear—but inaccurate (Figure 4.5). As I fill the bathtub, I am also watching the level of the water rise. As I follow the 'gap' between the water level and my 'desired water level'. I adjust the tap position to reduce the flow of water until I turn the tap off when the bathtub is full. In fact, I operate in a 'water regulation system' (Figure 4.6).

This system involves five variables: the desired water level; the bathtub's current water level; the gap between the two; the tap position and the water flow. This circle, or loop of cause–effect relationships is called a 'feedback process'. The process operates continuously to bring the water level to its desired position (Figure 4.7).

The feedback diagram tells the complete story of the action as opposed to the easier, but reductive 'cause and effect' expression.

When reading a feedback diagram, the main skill is to see the *story*: how the structure creates a particular pattern of behaviour. To follow the story, start at any element and watch the action develop. A good place to start is with the action taken by the decision maker.

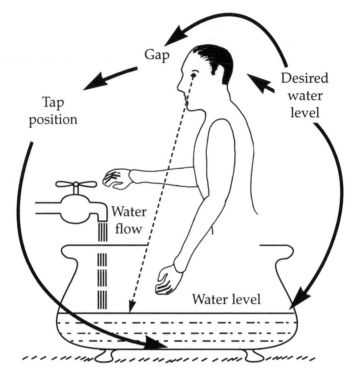

Figure 4.6 What really happens

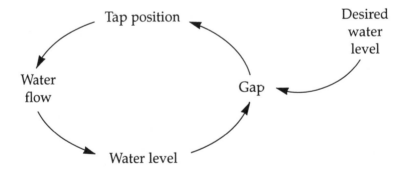

Figure 4.7 Influence diagram that underlies 'I fill the bath'

Delays

The wood-cutter example also highlights another important feature of systems: *delays*. The tree does not fall immediately, action has to be continued till a certain point where the tree will fall. The tree will fall slowly enough for the man to have time to yell 'Timber' and see the tree slowly crashing to the ground. In truth, in complex systems such as organizations, these delays are much longer. A typical organizational time-span is from six to twelve months between action and effect. If the man did not know within six months whether he had felled the tree or not, he would be likely to keep on axing far too much, probably cutting other things in the process as well.

> If I clap my hands loud enough, most people will jump. The irony of our lives is that we are genetically programmed to respond to immediate events. This, which was a real benefit for our ancestors—when confronted with a sabre-tooth tiger one had better react swiftly—is now causing us most of our grief. Our organizations deal for the most part with abstract, conceptual objects on a very different time-scale. Because change happens at more human time-spans, such as in quarterly or even monthly objectives, there is often a discrepancy between the strength of the action actually needed by the system and the actions prescribed by successive objectives probably creating violent fluctuations in the actual overall goal. The systems approach is dynamic in its very nature and as such is better adapted to a rapidly moving environment than most traditional approaches.

The concept of feedback allows us to link causal structure to dynamic behaviour. If a system shows a consistent pattern of behaviour, such as oscillations, goal seeking, start and go, or more simply, failure to grow, feedback loops enable us to identify the structural reasons for that behaviour and to learn how to act upon the feedback structures to modify that behaviour. Furthermore, the concept of feedback can highlight how the system itself causes its own behaviour: how reordering policies create fluctuations in inventories; how advertising wars are caused not by one company, but by several players; how, when using a common limited resource, the frantic optimization of the individual actors can swiftly lead to complete depletion of this resource for a ridiculously small gain. Understanding how the very causes of our problems may be in-built in the system is the first step to solving them.

By seeing whole systems, one can start to understand how the system causes its own behaviour. Where classical thought looks for causes 'out there', for external elements that might be responsible for whatever problem we have, and from then on allocates blame around, the systems approach investigates how the changing relationships between the elements of the whole system might be causing the observed behaviour. To start with, we can get rid of the 'blame' idea. There is no single cause, so there can be no attributable blame. We are either all guilty, or not at all. At this point, we can stop arguing about whose fault it is, and get on with solving the problem. At the International Rio de Janeiro conference on planetary environmental problems, one journalist described chiefs of states as men in a sinking ship wanting to know whose fault it is before they start bailing the water out. Furthermore, if a system is the source of a problem, it is also the mechanism for a solution. *By modifying the relationships between the elements of the system we are able to have a significant influence over its behaviour. In other words, 'change the rules, change the game'.*

Influences represent the actions in the system

The relationships in a system can be represented by the influences the components will have on each other: influences can be reinforcing (i.e. 'positive'), balancing (i.e. 'negative') or have lagged effects (i.e. delays).

Positive influences

'Positive' does not mean 'good', it just means that the influence reinforces the action: A $\xrightarrow{+}$ B means that quantities A and B vary in the same direction (if A increases, B increases—if A decreases, B decreases) (Figure 4.8).

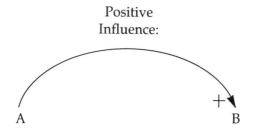

Positive
Influence:

A   B

Figure 4.8 If A increases, B increases. If A decreases, B decreases

Negative influences

A $\xrightarrow{-}$ B means that A and B vary in opposite directions (if A increases, B decreases—if A decreases, B increases). This is a

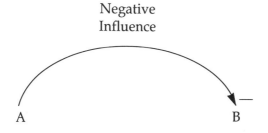

Negative
Influence

Figure 4.9 If A increases, B decreases. If A decreases, B increases

balancing influence, it will run contrary to the growth of the variable
it influences (Figure 4.9).

Delays

Finally, certain influences have a lagged effect. For instance, taking
aspirin affects the pain only after a few minutes. In this case the
influence between aspirin and pain will be delayed (Figure 4.10).

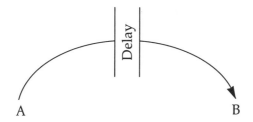

Figure 4.10 A has a delayed effect on B

Time plays a very large part in systems thinking, and is usually
represented by delays in different parts of the system between an
action and its effects.

Population dynamics

We can use the language of systems thinking to represent one of our
worst problems: population dynamics. Just as a bathtub that is being
filled by an inflow of water from the tap, and at the same time
emptied through an open plug hole, population is increased by the
number of babies born each year (Figure 4.11).

Figure 4.11 As births increase, population increases

Population is decreased by the number of people who die each year (Figure 4.12).

Figure 4.12 As deaths increase, population decreases

But the number of babies is not constant, it is influenced by the size of the population. Quite simply, the more people, the more babies. There is a *delay* of about one 'generation' between increasing the size of the population and the number of babies born. We can represent this system as shown in Figure 4.13.

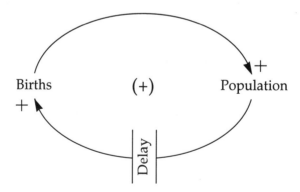

Figure 4.13 As population increases, births increase

This is a *reinforcing—or positive—loop*: the more people, the more babies; the more babies, the more people and so on ... population explosion. This reinforcing process is balanced by death: the more

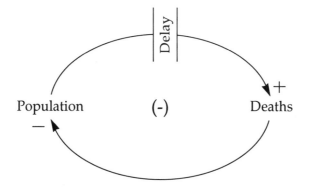

Figure 4.14 As population increases, deaths increase

people, the more death occurs, the more death, the fewer people. This is a *negative loop* (Figure 4.14).

Put together, we can represent the population dynamics system as shown in Figure 4.15.

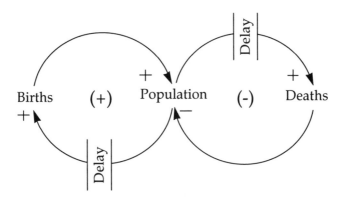

Figure 4.15 The structure of population dynamics

The same structural diagram can be applied to many different systems. If their structure is the same, their behaviour will be similar. For instance, the population dynamics structure equally applies to capital growth (Figure 4.16).

The same structure can be applied to the growth of knowledge in an organization: its learning capacity (Figure 4.17).

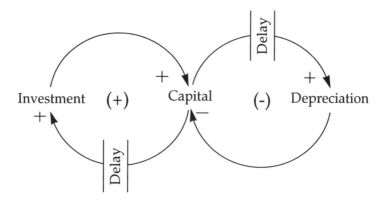

Figure 4.16 Capital growth

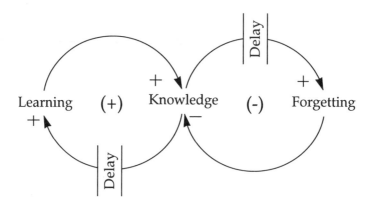

Figure 4.17 The dynamics of organizational learning and forgetting

Structure drives behaviour

The most powerful aspect of the feedback concept, is the way *we can link causal structure to dynamic behaviour*. This enables us to explain how sane people can be caught in irrational behaviour. They are part of a structure they do not see. For instance, if our company decides to increase advertising to attack a competitor's market share (Figure 4.18), it effectively pushes our competitor to launch an even bigger advertising campaign to stay competitive (Figure 4.19).

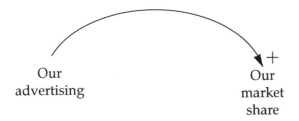

Figure 4.18 We increase our market share with more advertising

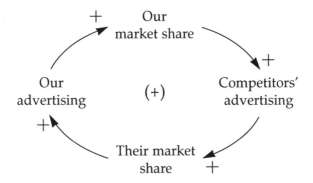

Figure 4.19 Advertising wars

This positive feedback loop is a vicious circle, an 'advertising war' from which neither competitor will benefit in the long run, with a lot of mind-bogglingly dull and expensive advertising for everyone to deal with. In this case, *the system causes its own behaviour*, our competitor perceives that we 'cause' advertising budgets to soar and vice versa, but one could equally claim that we create our own problem by stimulating the advertising build-up of our competitor. The relationships in this system make the advertising war inevitable, and we feel as helpless as our competitor (until we decide to redesign the system). As one of the original developers of the General Systems Theory, L. von Bertalanffy, writes:

> For example, the law of exponential growth is applicable to very different phenomena, from radio-active decay to the extinction of human populations with insufficient reproduction. This, however, is so because the formula is one of the simplest differential equations, and can therefore be applied to quite different things.[2]

Influence diagrams

Unless we have the right vocabulary to describe something, we tend not to perceive it. Because of the way our thinking works, what we do not have a name for tends not to be taken into consideration. Influence diagrams enable us to describe dynamic behaviour, and hence to highlight the underlying structure of events. For instance, if you take a small child to the park and place her in front of a swing, her natural reaction will be to push the swing, and to be very puzzled when the swing comes back to hit her. If you take her away a few steps and show her the whole structure of the swing, she will understand how it works, and the next time, move out of the way.

> A personnel department in a large company was complaining that the IT department could never implement a system properly. Whenever a new system was implemented, it would almost always crash. In the opinion of the users, the IT department never took the time to test its new systems properly and consequently they would fall over at the first opportunity. After discussing the reasons for this lack of testing, the personnel people realized that the reason IT did not test systems properly was because the department was under very tight deadlines set by personnel. The deadlines were set because of the bi-annual performance review, and personnel needed the latest system up and running before conducting the reviews. Unfortunately, they would ask for system upgrades precisely when they felt that the review procedure was coming forward. From seeing how they were creating their own problem, Personnel set a new rule for themselves: they would never ask for a change of systems less than three months before the performance review. Until they had drawn the loops, they did not even see what the problem was.

An influence diagram tells a story, much as our language does—with the additional feature of feedback loops. Influence diagrams can be thought of as sentences, which taken together can create a script of the dynamic evolution of the system. It is important to keep this idea of writing a story in mind when drawing out an influence diagram because most people tend to get carried away and link everything to everything else. We generally find it good practice to draw a maximum of three or four loops per diagram, and aggregate them later if necessary.

A language to describe and predict

Systems thinking provides a language to describe and predict the dynamic behaviour of complex systems. Our main reason to study systems is that their behaviour is predictable. We are trying to understand how past actions or structures influence the present. In many ways we have the same concerns as the ancient Greek tragic authors: trying to represent how past, and therefore hidden, actions mercilessly determine the outcomes of our present actions according to the rules of 'fate'. However, we believe that 'fate' can be modified by the actors if they understand how their system works. To quote E. F. Schumacher, the author of *Small is Beautiful*:

> In principle, everything which is immune to the intrusion of human freedom, like movement of the stars, is predictable, and everything subject to this intrusion is unpredictable. Does that mean that all human actions are unpredictable? No, because most people, most of the time, make no use of their freedom and act purely mechanically. Experience shows that when we are dealing with large numbers of people, many aspects of their behaviour are indeed predictable; for out of a large number, at any one time, only a tiny minority are using their power of freedom, and they often do not significantly affect the total outcome.[3]

In a way, our entire concept of 'organizations' is there to limit individual freedom—to a few people at the top. Yet, because of the structural complexities of organizations, even the organizers get lost. Systems thinking gives us the tool to represent the dynamics of such complex systems—and hence, to understand them better.

Operational thinking

Systems thinking provides a vocabulary to describe organizations in *operational* terms. Mostly, we tend to think *correlationally* rather than *operationally*. Barry Richmond describes the difference between operational and correlational thinking. *It is a matter of cows!*

> A recent article in a prestigious economics journal reported on a model destined to forecast milk production. The model used statistical methods to distil a set of correlations between yearly milk production and movements in a series of macroeconomic variables (things like GNP, interest rates, agricultural commodity prices and inflation). When this set of macroeconomic variables was blended together using appropriate statistical weightings, the resulting model tracked the volume of historical milk production very closely.

> Such a model is just ducky if your purpose is to forecast next year's milk production volume—ducky, that is, if next year turns out to be pretty much like this year and last. However, such a model would be of little value if things changed, or if your objective was to figure out how to

increase milk production. If we relied on a correlational model for guidance in these cases, we would be led astray. We'd be led astray because we were thinking correlationally rather than operationally. The fact is, the things like GNP and interest rates have nothing whatsoever to do with the actual 'physics' of milk production. The physics of milk production?! Yes, the 'physics' of milk production! COWS produce milk, each one of them yielding a certain amount per unit of time. No cows, no milk production. Basic barnyard logic. True, it's not as elegant as some macroeconomic theory, but it is far less abstract—and hard to argue against.[4]

Systems components

Jay Forrester has profoundly influenced the field of systems dynamics (systems simulations assisted by computers), and set the main methodological rules to the modelling of systems on computers. In an interview with a consultant, he posits that:

> There are two—and only two—kind of variables in a system dynamics model: levels (or accumulations) and rates (or actions). Once you believe that there are only two kinds of concepts in a system, everything you look at has to be one or the other. The idea of two and only two kinds of variables should not be new to managers. Accounting reports are cleanly divided between the balance sheet and the profit and loss statement. Balance sheet variables are system levels; profit and loss variables are system rates, which causes the balance sheet levels to change.[5]

In operational terms, the main elements we need to focus on are resources, actions, motivations, conditions and delays. In systems dynamics models, resources can be modelled as accumulations and actions as flow rates. Conditions will be represented as influences.

Resources

Aristotle said: 'Ex nihilo nihil fit'—nothing comes from nothing. This fundamental insight has dogged science and philosophy for decades. Nothing that we do or produce comes from nothing. Products come from something, and go somewhere once they are consumed (i.e. destroyed), they do not simply disappear with the trash truck. Any activity to create uses up a resource. Some resources are plentiful, such as solar energy, or sea water, while some others are rare such as gold or uranium or goodwill. Resources are any stockpiles of elements we are going to use for action. They can be physical resources, such as oil, water, oxygen, materials; or more abstract like financial resources. They can also be very subjective like motivation or time—both can be used as a resource. Resources fall into two categories: non-renewable and renewable.

Non-renewable resources
Non-renewable resources are resources of which we have only a small amount or which are very slow to renew—such as oil or ore which can be used up and *have to be treated as capital, not income*. As E. F. Schumacher writes in *Small is Beautiful*:

> The illusion of unlimited powers, nourished by astonishing scientific and technological achievements, has produced the concurrent illusion of having solved the problem of production. The latter illusion is based on the failure to distinguish between income and capital where this distinction matters most. Every economist and businessman is familiar with the distinction, and applies it conscientiously and with considerable subtlety to all economic affairs—except where it really matters: namely, the irreplaceable capital which man has not made but simply found, and without which he can do nothing.[6]

Renewable resources
Certain resources are renewable over time. Financial resources, for instance, can be renewed through profitability (the concept of self-financing). Solar energy or time are also renewable resources which we can treat as income: we get the same amount of it every day and the choice of what we do with it is up to us. Some resources, however, can be deceptive. Oxygen, for instance, is an unlimited, renewable resource at an individual level: we cannot imagine running out of air to breathe. At a larger scale of communities or societies, even oxygen can become a non-renewable or rare resource.

> Mexico City is built in a mountainous hollow at about 2000 metres above sea level. In the winter, the cold air on the top of the hollow forms a lid from which the polluted air of the city cannot escape. Consequently, the air in Mexico City can become literally unbreathable, and several people die each year in the centre of asphyxiation. This particularly affects young children, who now have a special holiday from January to February, so that they can be taken into the countryside during the worst period of the year. Although the situation is obviously drastic, nothing can be done to regulate the millions of polluting cars and industries in a city that is now so vast it needs the transport and the jobs to survive.

Actions

Actions are what we *do*. In most cases, action uses resources. In systems thinking, we shall try to focus on rare resources that can be used up without our noticing it. Actions can be expressed mostly in terms of flows. Each action point can be represented by a decision centre. A decision centre receives information and transforms it into action. This, in corporations, is the role of the manager. Management can be seen as the art of transforming information into action. Information, both internal and external, arrives on the manager's desk, and he or she must then act upon it. The manager's action will change the situation and will generate more information and so forth.

> Forrester's approach to decision making is based on a simple feedback loop: Information provides a basis for decisions, which then turn into action, generating more information for future decision making, and so forth (Figure 4.20).

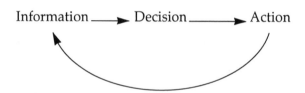

Figure 4.20 Transforming information into action

> In other words, management is the art of transforming decisions into actions. In more detail, he argued that decisions usually involve three components: perceived actual conditions, desired conditions and corrective actions. In making the difference between perceived and actual conditions, Forrester attempts to take into account the long delays involved in the transmission of information through organizational structures. Something may have already happened—but it cannot show for several months![7]

In that perspective, the largest action is not necessarily the most efficient. If we look at a problem with the *'Big is Important'* mentality we might focus on the low-payoff actions. For instance, if you bake bread, you must add yeast to the flour. Although a very small percentage of the mixture, without the yeast, the bread simply will not rise. Adding more flour will not help. If we are not aware of this, we may be led to consider that energy is not an important concern of a nation's government because it represents less than 10 per cent of its budget.[8] Similarly, most companies tend to consider that training is a 'secondary issue' often delegated to outsiders. Training, however, is probably the most influential activity on the firm's continued success and longevity.

Motivations

Motivations are the reasons that push people to do things. They may be personal, or may be imposed by the environment. In businesses, for most of the time motivations of departments are to a large extent imposed by the structure—rather than by the people. Whenever someone does something we disagree with—or something which is a problem to us—in most cases, our first reaction is to think 'what an idiot!' (or any other favourite insult that basically expresses that the other person is somewhat sub-human). Alternatively, we take people's action personally, even when they do not even know we exist. This emotional reaction is legitimate, but it can often cloud our best judgement. By way of contrast, thinker Edward de Bono talks about logic bubbles.[9] If we assume that the other person is behaving perfectly rationally in his or her context we unearth his or her 'logic bubble'. In this bubble, apparently 'idiotic' actions will make a lot of sense according to *that person's motivation*.

In systems thinking, we tend to assume that a system's goal is what it actually does as opposed to what it is supposed to do. By understanding the motivations of the elements, we can reconstitute the overall action; the problem is then one of consistency. It is perfectly reasonable to expect a sales department to sell as many products as it can, even if production cannot deliver.

> The management team of a US telecom company was concerned about the productivity of their customer service centre. They called in a team of productivity consultants who immediately set out to measure the number of incoming calls and the number of operators available to deal with these calls. They then calculated the average time taken to handle a customer and from this determined the ideal number of staff needed. The upshot of it was to reduce the number of staff. For this reduction to work, they had to keep the calls close to average time. Disregarding the fact that the average had been calculated by including both the short and the long calls, they set up a system by which the operator was penalized if he or she spent more time than average on the phone (a red warning light would flash up).
>
> Management went for the idea, and it took the company eight

months to realize that they were losing customers because of this policy. What happened was that the people with a difficult problem to handle would be cut off in mid-conversation, ring again and be picked up by a different operator. After a few such experiences, they would be annoyed enough to change phone companies. The productivity consultants had done what they were supposed to do: they improved productivity. Management was responsible for the confusion when they assumed that the productivity of the service centre could be improved without putting quality first.

Conditions

Conditions are what people monitor to see whether they have achieved their goal. The chosen conditions can have huge impact on the actions. For instance, if we monitor *share price* we will take different decisions from those we would take if we monitored *staff well-being*. In one case we will tend to push for short-term profits, to reassure investors and run the risk of mortgaging the future of the firm. In the other case, we might not be able to take ruthless decisions even though they are necessary when *the good of the many outweighs the good of the few*. Conditions are very important because they are our points of reference: how we know if we do well, or badly; how we know when to act some more, or to correct our action, etc. By analysing the conditions that systems monitor, we can understand how some groups can have actions with effects contrary to their motivations—and keep doing so.

> Politicians, for instance, monitor opinion polls rather than the real effect of their policies (which are in any case more difficult to monitor because of the complexity and delays involved). As a result, many policies will have contrary effects to those intended— but will be pursued as long as the 'general opinion is favourable'. The general public has even less means than the politicians to link effects to cause ... Furthermore, because politicians are well known for not keeping promises, it is more important to appear consistent than not! When a policy—such as rigid monetarism in recent years—starts to raise doubts about its actual outcome their answer will tend to be: 'It's not working because it hasn't been applied relentlessly enough!'

In practice, most managers are aware that *'what gets measured gets done'*. Yet, they seldom apply this insight rigorously in trying to identify what is implicitly measured by the system. Alternatively, they apply this directly by imposing goals on people with an 'or else ...' attitude. They often find that these goals are achieved—but at what cost? The resources necessary to achieve one particular goal are often diverted from somewhere else and unless managers are very aware of their real priorities, a pure 'what gets measured gets done' approach can lead to a series of inconsistent actions that increase overall confusion rather than aiding progress of the company as a whole.

Overview

The systems thinking framework uses a number of conceptual tools to represent and describe reality according to this new set of assumptions. Principally, this means drawing out feedback loops by expressing influences and delays. Systems theory focuses on certain 'operational' aspects of situations. These aspects are mainly about resources, renewable or non-renewable; actions; motivations; and conditions. Put together, these elements will help us to draw out systems control loops. These elements are the aspects of a situation a 'systems thinker' will try to draw out at the beginning of any analysis, or action: they are the key to understanding how systems behave.

Notes

1. 'Making acquisitions work: lessons from companies' successes and mistakes', *Business International*, 1988, Geneva.
2. L. von Bertalanffy, *General Systems Theory, Foundations, Development, Applications*, 1968, George Braziller.
3. E. F. Schumacher, *Small is Beautiful*, 1973, Blonds & Briggs.
4. Barry Richmond, *Ithink User Guide*, 1990, High Performance Systems.
5. In M. Keough and A. Doman, 'The CEO as organisation designer', *The McKinsey Quarterly*, 1992, No. 2.
6. E. F. Schumacher, *Small is Beautiful*, 1973, Blonds & Briggs.
7. G. Richardson, *Feedback Thought in Social Sciences and Systems Theory*, 1991, University of Pennsylvania Press.
8. For an insightful discussion of this aspect of systems, see: D. Meadows and J. Richardson, *Groping in the Dark*, 1980 John Wiley & Sons.
9. E. de Bono, *Thinking in Action: the de Bono Thinking Kit*, BBC Education and Training.

Chapter 5

Basic systems behaviours

The mathematical basis of the general systems theory shows that when a system exists in a simple configuration it will behave in certain specific ways such as exponential growth, oscillations or cyclical behaviour.[1] These equations, originally modelled by Maxwell have been observed in practice by researchers of many various field—from the behaviour of machines, electrical circuits, predator–prey populations, weather patterns, etc. As far as we know, they are the best explanatory representation of dynamic systems. In this chapter, those basic systems behaviours will be briefly presented.[2]

- **Exponential growth**. When a system is driven by a dominant positive feedback loop, it will produce exponential growth (or decay) (Figure 5.1).

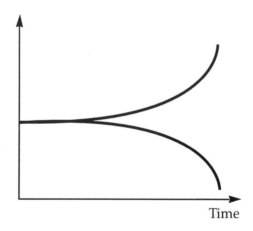

Figure 5.1 Exponential growth or decay

- **Goal seeking**. A tendency to return to original state after a disturbance indicates at least one strong negative feedback loop (Figure 5.2).

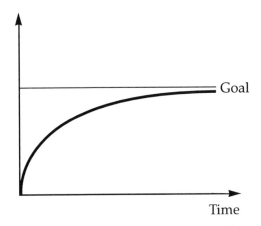

Figure 5.2 Goal seeking

- **Oscillations**. A negative feedback loop with a delay in it can produce oscillatory behaviour (Figure 5.3).

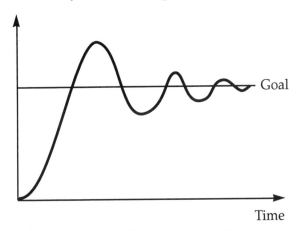

Figure 5.3 Oscillations and stabilization

- **S-curves**. S-shaped growth results from linked positive and negative loops that respond to each other (Figure 5.4).

We must note that, at this stage, systems thinking makes its most important assumption: any of the previous behaviours are linked to a feedback structure. It is easy to verify that if we create a certain feedback structure then we will observe the expected behaviour. This has been done often enough for us to accept the validity of the causal link. It is, however, more difficult to prove the reciprocity: that if we observe such behaviour, we can trace it to an underlying

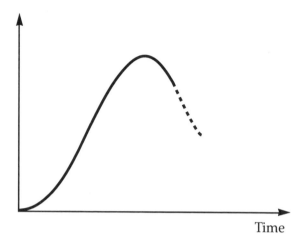

Figure 5.4 S-curve: the shape of growth

feedback structure. Systems thinking assumes this to be true, and indeed some of these feedback structures can be exposed.

The fundamental insight is that actions do not exist in a vacuum. Although we tend to think of actions *in themselves* as disconnected from the entire context, in reality they are not. Any action will stem from certain conditions and have an impact on the *whole context of the conditions*. This will either reinforce the conditions for action—and thus the action—or balance them out. In effect, we see a feedback structure appearing from the simple truth that actions cannot exist out of context, i.e. in isolation. In the following pages, we shall try to show how the interaction between actions and their context (conditions) can generate our fundamental behaviours of exponential growth or decay, goal-seeking behaviour, oscillations or S-curves.

Positive feedback loops: the engine of growth

The old sage offered the emperor a simple bet; although the emperor is akin to the gods themselves, he will never be able to place on a chessboard the following grains of rice: one grain of rice on the first square of the chessboard; two on the next square; four on the following; eight on the fourth square and so forth ... until the sixty-fourth square. The emperor laughed and offered the old sage his weight in gold if he won and death if he lost. He then called for a chessboard and some rice. As can be expected, the emperor will never be able to place all these grains of rice on the chess board, because they follow an exponential growth law: for square n, the number of grains needed would be 2 to the power $n - 1$. The sixty-fourth square would then necessitate 2 to the power 63 grains, that is in the order of 9 exp 10 at the power 18, more than nine million billions.

A quantity grows exponentially when it doubles itself at regular intervals. Take a piece of paper and fold it in half. You have just doubled its thickness. Fold it in half again to make it four times its original thickness. Supposing we could go on folding the paper like so for a total of 40 times, how thick do you think it would get? In fact, it cannot be done. But if the thickness of the paper could be doubled 40 times over, it would make a pile of paper high enough to reach the moon !

Exponential growth is often the main driver of a system. At the same time, in most natural systems, the system itself controls itself strongly in order to prevent exponential growth. It is particularly relevant to us because as opposed to most animal populations, the human population follows at the moment a law of unchecked exponential growth (Figure 5.5).

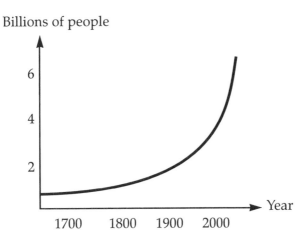

Figure 5.5 Population growth

In systems thinking terms, exponential growth or decay is generated by a positive loop. Positive loops are more commonly known as virtuous circle or vicious cycles. The positive loop is a self-reinforcing process where the elements 'feed' off each other; the growing action reinforces the condition for growth which in turn generates more growing action and so on. With exponential growth, the increase is proportional to what is already there: if a child puts £10 a year in a piggy bank, his or her savings will increase in linear fashion, i.e. the amount of growth is constant. However, if the child invested £10 at 7 per cent per year interest—and let the interest accumulate in the account—the invested money would grow exponentially. None the less, we tend to anticipate linear growth. Suppose you own a pond on which a water lily is growing. If the plant were allowed to grow unchecked, it would completely cover the pond in 30 days, yet, on the 29th day, the plant only covers half the pond and on the 25th day only 1/32nd of the pond.

According to Jay Forrester the growth of the systems dynamics field is a natural consequence of exponential growth:

> Suppose the number of system dynamics practitioners doubles every three years. If you start with one person, in three years there will be two; in six years, four; and so on. In the 36 years since systems dynamics began in 1956, there has been time for 12 doublings. Such exponential growth would bring the number today to over 4000. Probably at least that many are now engaged in one or another aspect of system dynamics.[3]

Vicious and virtuous circles

A positive feedback loop is what we know as a 'vicious' or 'virtuous' circle. In fact, positive loops can be seen as amplifying divergence. An action (or group of actions) will modify its context in a way that *reinforces* the conditions favourable to that action. The action will then be exerted with more force, thus increasing the conditions again, then the action again and so on. Positive feedback loops can be generically described by the influence diagram that comprises Figure 5.6.

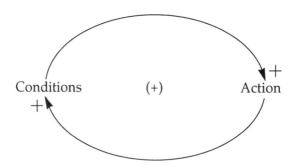

Figure 5.6 A reinforcing process: the positive feedback loop

This is a self-reinforcing process, what we otherwise call a virtuous or vicious circle. The action increases the conditions which in turn reinforce the action. This process feeds on itself to become ultimately explosive (doubling the quantities regularly). There is nothing 'positive' (as in good) about it, positive is simply a convention that describes the fact that the feedback influences are positive all around in the diagram. In fact, most of the time, positive feedback is quite unpleasant and found in the likes of: spread of a virus through a population; panic in a crowd; hurricanes; or the strident sound you get if you place a microphone too close to the amplifier.

Furthermore, a positive feedback loop can turn from a virtuous to a vicious circle: it can reverse itself, and the same mechanism that has caused explosive growth can cause extinction. These positive loops are very hard to break and can at times defeat the best efforts to modify them.

> A classic example of a vicious circle may be found in British industrial relations. Over the years, Britain has settled into what academics call a 'low-skill equilibrium'. Through promoting a free job market, the government has made it easier for companies to poach skilled employees from their competitors rather than training them themselves. Consequently, most companies become very reluctant to train because they figure that if they invest in their workers, they will make them attractive to rival companies who would them poach them by offering higher wages. After a few years of this attitude, no real training is being done, and the general skills level of the country is going down. Considering the difficulties involved in rebuilding a skilled workforce, the attitude is then to focus on low-value-added products, and to compete on price. This low positioning reinforces the low-skill equilibrium further.

Sales, word of mouth and reputation
A good example of a virtuous circle is the way more sales mean more satisfied customers, which generate a positive word of mouth, which in turn boosts sales, to create more word of mouth and so forth (Figure 5.7).

This is also how reputations, good or bad, build up. Rumours spread on this principle and can lead to total panic. The truth is that when things go well, several positive loops can be coupled and together create this groovy feeling that all is for the best.[4]

The presence of a reinforcing loop does not mean that a population

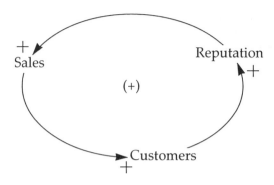

Figure 5.7 The bandwagon effect

will necessarily grow exponentially; it only means that it has the structural capacity to grow. Biologists have found that when an eco-system is disturbed by an outside influence, one of the constituent populations will tend to take advantage of the change and grow exponentially—until it further modifies the system by its growth and then needs to settle down if it wants to survive.[5] The positive feedback structure was in place, but needed some room to develop.

> The immediate consequence of the American deregulation of airline companies was an explosion of small local companies (some of which went on to be famous, such as People's Express). These small companies thrived for a while, until they started both encountering internal problems due to over rapid growth and competing against themselves. After a few years, the decks had cleared and the unexpected outcome of the deregulation is that the American air travel industry is now more concentrated than ever in the hands of a few key players.

Is growth good?
Exponential growth can pose severe difficulties for companies. The feeling that you can simply grow sales, and the rest will follow fails to take into account the systemic implications of growth.

> Gregory Bateson tells the tale of the polyploid horse: a brilliant geneticist fiddles with the DNA structure of the common cart horse (*Equus caballus*) and creates a horse precisely twice the size of the ordinary Clydesdale. It is twice as long, twice as high, and twice as thick. It is a polyploid, with four times the usual number of chromosomes.

> The scientific community applauds this achievement, but unfortunately, the creature cannot stand on its legs because they would snap under its weight. It needs to be constantly showered

with water because the beast's dermal fat is twice as thick as normal and its surface area only four times that of a normal horse so that it does not cool properly and its insides are cooking. It is also constantly panting, to oxygenate its eight-times body with a windpipe, after all, only four times the normal area of its cross-section.

And then, there is eating. Somehow, it has to eat, every day, eight times the amount that would satisfy a normal horse and has to push all that food down an oesophagus only four times the normal size. The blood vessels, too, were reduced in relative size, and this makes circulation more and more difficult and puts extra strain on the heart. As Bateson puts it: 'a sad beast'.[6]

The tale of the polyploid horse is relevant to many companies who grow exponentially without bothering (or knowing) how to put the appropriate support structure in place. Two main dangers face these companies: either they are cash-rich and they outgrow support, creating huge administrations to administrate their product; or they disappear as fast as they appeared through an incapacity to sustain their success because the relevant structures are not in place.

Negative feedback loops: the source of stability

Positive feedback loops, however, seldom exist unchecked for long. Everything, everywhere, is in constant change. Any system that is going to survive long enough for us to take any notice must be able to cope with change. All stable systems deal with change by acting to cancel or negate any change. This is called a negative—or balancing—feedback loop. Negative feedback loops tend to hold a system within some acceptable range or return it to a stable state. A negative feedback loop will spread the consequences of change, until it creates a reaction that will come back to limit the effects of change.

For example, water is essential to life and is constantly being used to flush wastes out of the system, and for many plants and animals, help with cooling down. When the water content of our body drops beyond a certain level, we feel *thirsty* and adjust that feeling by drinking water (Figure 5.8).

Plus c'est la même chose, plus ça change. In fact, most systems are self-corrective. As we have seen, positive feedback tends not to be appreciated by the system as a whole and is likely to trigger one or several negative feedback loops. In most systems, one variable is maintained as steady as possible by changing all the other variables.

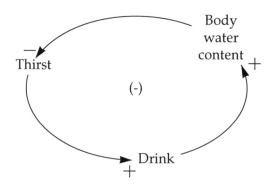

Figure 5.8 Regulating the body's water content

This variable can be survival for populations, temperature for a furnace, or Faith for the Church. *This variable can be considered to be the purpose of the system.* However, in human designed systems, several confusions can occur. For instance, most management scientists would agree that survival is the purpose of most companies. In practice, companies do not know very well how to ensure their survival, but their board knows where its chips are placed. Consequently, the variable the company will try to maintain is share price. This, in some instances, can go against the survival of the firm.

A balancing or goal-seeking process

In balancing processes, the action will tend to 'correct' the conditions which triggered it in the first place. The more one acts, the less the conditions appear, leading to a smaller and smaller action— until it stops completely. These balancing processes are described by a negative loop (again, there is nothing 'negative' about them); i.e. a loop where there is an odd number of negative influences (when one element grows, the other decreases, etc.) (Figure 5.9).

In a balancing process, the gap between the condition and the goal, or limit, motivates the action. A simple instance of such a process is how someone adjusts the temperature in a furnace by adding coal— or letting it burn low. The discomfort of the room being either too cold or too hot expresses a gap between a comfortable state, a desired room temperature, and the reality of the actual temperature in the room. If the gap is noticeable, the person will try to reduce it by either shovelling coal in the furnace, or letting it burn down (Figures 5.10 and 5.11).

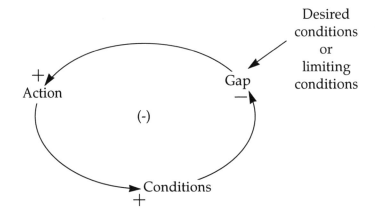

Figure 5.9 A balancing process: the negative feedback loop

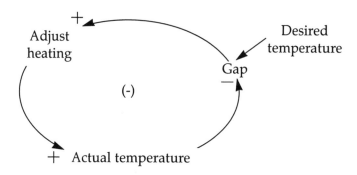

Figure 5.10 Regulating a coal furnace

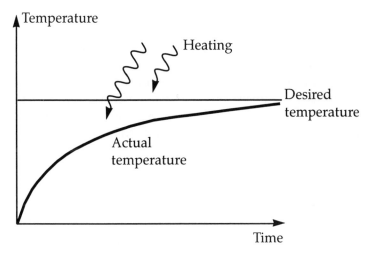

Figure 5.11 Graph of regulating a furnace

The thermostat

Most negative feedback loops need some sort of 'thermostat' to regulate the system. In the previous furnace example, the person in the room was playing the role of thermostat since he or she decided whether or not to stoke the furnace. One of the most common mechanical systems is the heating system found in most homes and buildings: once the temperature has been set on the thermostat (goal), the system will try to keep the temperature in the house at that level. In this case, a mechanical device is used to keep the furnace going at a 'desired' temperature (Figure 5.12).

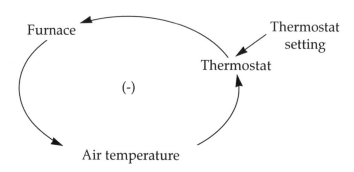

Figure 5.12 The thermostat

Oddly enough, when we feel that a room is too warm, our immediate instinct is to open the window. This lamentable non-systemic approach will work for as long as the thermostat needs to think 'aha!, the air is getting colder—I'd better do something about it.' and turns the furnace on. Consequently, in a few minutes the room is back at its normal temperature but with a higher energy cost and with an open window. This tendency to express a problem as a solution and then to jump to conclusions—and actions—is very widespread in business situations. Too often, the 'Don't just do something, stand there!' injunction is disregarded and solutions create more problems that need yet more solutions with disastrous effects on the overall outcomes. When you add a 'what gets measured gets done' attitude to this tendency, you can end up with absurd—and costly—crises.

A high-tech pump manufacturer measured the production of its company by the volume of pumps shipped each month. Consequently, there was a drive to rush as many pumps as possible to the customer at the end of each month. This led to pumps lying around at the customer's premises waiting for additional parts or the installation team.

The sales department then intervened to try to maintain the relationship with its customers by storing the pumps out of the factory in its own premises. This turned out to be difficult and costly since there was not enough storage space (why would the sales department need it in the first place?) Furthermore, the plant triggered invoicing on dispatch, so now the customers would receive invoices for pumps they had never seen. The whole mess was finally sorted out by changing the measurement system at the plant and an agreement to let sales decide on the dispatching. Unexpectedly, this led to a closer relationship between the sales department and the factory, better information about demand and less inventory of finished goods.

In this case, the 'thermostat' was set by the measuring system of the plants: so many units out per month. The sales department reacted to the problem by 'opening the window rather than changing the thermostat' and its solution only compounded the problem.

In a balancing process, the crucial element gradually adjusts to its desired level. Organizations and societies resemble complex organisms in that they have a myriad of balancing feedback processes. In companies, the production and materials ordering process is constantly adjusting in response to incoming orders: short-term (discounts) and long-term (list) prices adjust in response to changes in demand or competitors' prices; borrowing adjusts with changes in cash balance or financing needs.

The expectancy theory of motivation works on the principle of a negative feedback loop: aspiration is defined as the target, or goal, we set ourselves. It is what we really intend to achieve. Consequently, people who achieve most are those who set themselves ambitious but realistic targets (Figure 5.13).

The perceived gap between reality and expectations or aspirations motivates us to work harder, which in its turn increases the perceived professional success and helps us to settle at the aspired level by reducing the gap. In a specially designed experiment involving 120 professional negotiators from the aerospace industry, US investigator Chester Karass found that:

(1) high aspirers got better deals; (2) they always won against low aspirers (even when the low aspirer had greater power and skill); and (3) skilled negotiators with high aspirations were winners even when they had less power.[7]

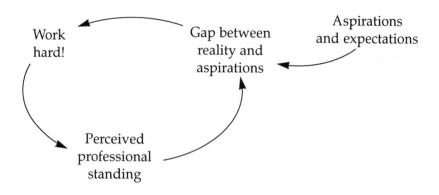

Figure 5.13 High aspirers get better results

Supply and demand: two negative loops coupled together

The law of supply and demand, which ideally determines market price, is usually represented by economists as shown in Figure 5.14.

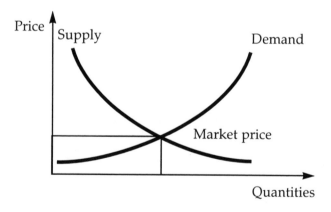

Figure 5.14 The law of supply and demand

As demand increases, prices goes up. As supply increases, prices come down. We can represent this system as two balancing loops linked by price (Figure 5.15).

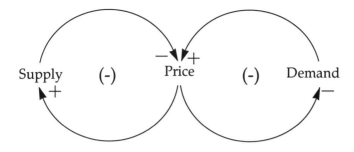

Figure 5.15 Price is determined by two negative feedback loops

With supply, price and demand unconstrained, the system should stay stable: it should settle at the 'right price'. However, in real life this system has considerable delays and anticipations due to the individual response time and strategies of the players.

> For instance, at the event of the Gulf War, some 'experts' asserted that since there was a glut in the oil market, the price of the barrel should not increase by much. In truth it rocketed up as had been projected by most people, for two main reasons: on one hand a panic effect that made demand go up; and on the other hand a considerable delay before the potential added capacity could actually come 'on line' and service the demand.

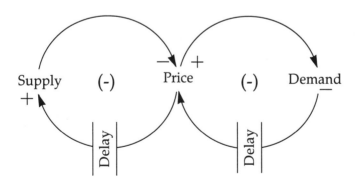

Figure 5.16 Supply, price and demand unconstrained

In most cases, not taking delays into account can lead to skewed expectations. Our intuitive understanding of stock and flows is not very developed and we tend to consider quantities as a whole—ignoring the fact that they can accumulate, or deplete, over time. If long delays are involved, we must think very seriously about the overall quantities involved and the flow rates at which they will appear.

Cycles are created by delays in negative feedback loops

Daily life is full of small and not so small overshoots. A car on an icy road can slide past a stop sign. If you eat or drink too fast, you can go too far before your body sends unmistakable signals that you should stop. On a larger scale, a fishing fleet can become so large and efficient that it depletes the fish population upon which it depends. Developers can put up more buildings than people are able or willing to buy, and so forth.

Delays between consequences and actions are everywhere in human systems. Lately, management theory has started to focus on delays in production systems. One way of explaining Japanese marketing success is by showing that their development delay is about 50 per cent shorter than that of their competitors, which enables them to bring out new cars much sooner, and to keep closer to demand. Delays in a system are a fact of life. One can try to shorten them, thus gaining responsiveness for the system but the most important thing is to be aware of delays: how long, and where. Virtually all feedback processes have some form of delay, but often these delays are either unrecognized or misunderstood. *This will typically result in overshoot*, going further than needed to achieve a desired result. Unrecognized delays can also lead to instability and breakdown, especially when they are long.

Negative feedback loop with delay
With a goal in mind, an actor will adjust his or her behaviour relatively to a condition that tells him or her 'where he is' relative to the goal. A delay between the conditions and the real position of the system causes overshoot, where the action is pursued further than necessary and passes the mark (Figure 5.17).

This is a balancing process, with a delay between the action and its effect on the conditions. The system is going to oscillate till it reaches equilibrium. For instance, if you drive a car on an icy road, there will be a delay between the moment you press the brakes and when the car stops. Unless you expect this delay to be far longer than normal, you are likely to brake later rather than sooner, and to overshoot the mark. Crash, bang!

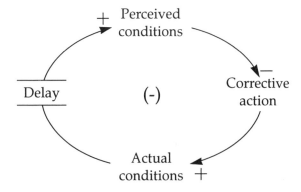

Figure 5.17 Negative feedback loop with delay

Ye olde shower: a balancing process with delay
Have you ever taken a shower in an unfamiliar bathroom. If the system is ancient, there is a delay between the action of turning on the tap and the effect on the water flow (Figure 5.18). Consequently, you step in and turn the hot water on. The water remains cold. So, you persist and turn up the heat some more. When water arrives, it is scalding hot. You swear, jump away and quickly turn the tap back to cold water. In a few seconds, it is freezing cold again, and so on ...[8]

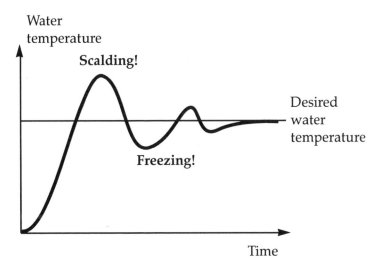

Figure 5.18 Ye olde shower

You are in fact suffering from a long delay between the action of turning on the tap and the arrival of the water. If, by turning up the heat, the water remains cold, we tend to perceive that our action had no effect. So we act further. In fact, the effects of our action appear with *a delay*. Each cycle of adjustment will compensate somewhat for the cycle before: This can be represented by the influence diagram (Figure 5.19).

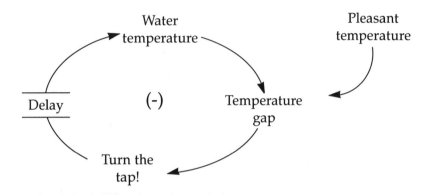

Figure 5.19 We change the tap setting to get the water at a temperature we like ...

The more aggressive we are in our behaviour—the more drastically I turn the tap—the longer it will take to reach the right temperature. Often by being pro-active, you can in fact increase the instability in the system. Aggressiveness tends to produce instability and oscillation instead of moving quickly towards our goal.

In a systems thinking seminar, I once ran the following experiment. By using a funnel driven in a shoe box, I would roll golf balls towards a target. The result was a spread of positions around the target (Figure 5.20).

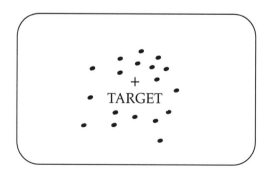

Figure 5.20 Target practice

Considering that the result was not good enough, I then started to adjust the position of the funnel according to where the last golf ball had landed relative to the target. If it is right of the target, I move my funnel left, and so on. The surprising result was an even wider spread around the target (Figure 5.21).

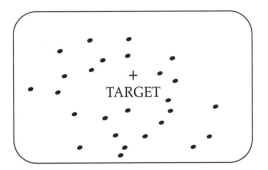

Figure 5.21 Target practice

A sales manager reviews the sales figures for her group. She is pleased to see that almost half her staff performed above average last year. However, about half are below average and there are a few underperformers in the bottom 15 per cent who have not done very well.

To reward excellent performance and help improve the worse performances, she sets up the following incentive scheme: anyone selling above a certain target level will earn a bonus. Anyone below a certain level will be penalized. Is she likely to improve the performance of her department?

In this case, we were introducing a feedback link where it was not needed. This apparent corrective mechanism in fact introduces even more spread by reinforcing divergences. This frequently happens when people try to control systems too tightly. They introduce so many corrective mechanisms that sooner or later they inadvertently put positive loops in place that will amplify divergences rather than balance them as intended.

Production capacity

Production capacity is hardly ever at the right level because of the delay involved in the investment decision. Because the investment decision is so hard to make, it is often left as late as possible. However, there is such a delay between when the decision is made and when the new capacity comes on line, that the investment will

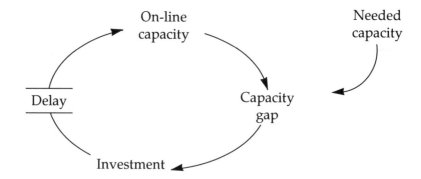

Figure 5.22 Watch the delay between the investment decision and the increased capacity

tend to be over-estimated. Where the capacity is operational, we are likely to see over-capacity (Figure 5.22).

Capacity can be linked to demand. For instance, if growth approaches a limit which can be pushed into the future if the firm invests in additional capacity, the investment must be sufficiently aggressive and fast to forestall reduced growth or else it will never

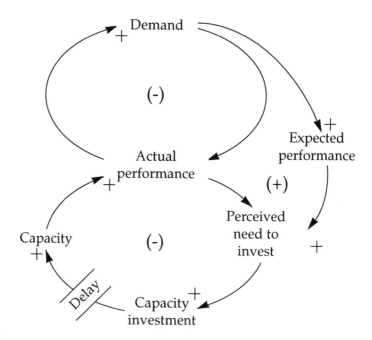

Figure 5.23 We need to invest before the increase in demand outruns our capacity and erodes our production standards

be made. Growth will be reduced by the effect that working with under-investment will have on the performance of the firm (Figure 5.23). When this happens there is a self-fulfilling prophecy where lower goals lead to lower expectations, which are then borne out by poor performance caused by under-investment.[9]

Growth: an S-curve process

The organization of every complex system is built from the same two simple elements: positive and negative feedback loops. When put together, they generate S-curve behaviour.

If we consider again the population structure we had earlier and apply it to, say, a fish population, we can see how the positive loop can drive exponential growth. If each fish spawns on average five other fish, the more fish there are, the more will be born and the larger will be the population. This can be represented by the very simple equation:

Population $(t+1) = 5 \times$ Population $(t) +$ Population (t)

However, as the number of fish increases, the number that die off each year increases also—this is the negative balancing loop. Supposing that a fish has an average life span of 3 years, we can approximate this equation by saying that one-third of the population dies off each year:

Population $(t+1) =$ Population $(t) -$ Population $(t)/3$

By combining the two expressions, we get an equation that describes the evolution of the fish population.

Population $(t+1) =$ Population $(t) + 5 \times$ Population (t)
$-$ Population $(t)/3$

which can be written in its more general form as:

Population $(t+1) =$ Population $(t) +$ Birth rate $(t) \times$ Population (t)
$-$ Death Rate $(t) \times$ Population (t)

and can be represented by the influence diagram in Figure 5.24.

In our example

Population $(t+1) = 5.6 \times$ Population (t)

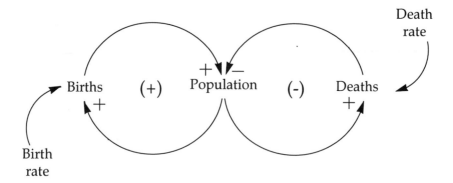

Figure 5.24 Population dynamics

In this case, the population will still increase exponentially. However, if this fish population live in a contained area, sooner or later they will 'overgraze their range': there will be so many fish that they will starve because their natural milieu cannot support such large numbers of fish. The death rate is likely to increase dramatically, so that from exponential growth, the growth of the population is likely to slow down, and can even reverse itself to start an exponential decline. Unless the milieu can reconstitute itself fast enough (that is before all the fish have died) then, as soon as there are fewer fish, the milieu (algae, plankton, etc.) will itself grow exponentially, making room for the fish to go through another cycle, and so on.

Interestingly enough, this is exactly what happens to fishermen. They tend to fish a species so extensively that they drive it to near extinction. Hopefully, they are then forced to stop (because the fish cannot be found any more) which enables the population to rebuild. Unfortunately, in some cases we have been so thorough in our search for the last of the species that we have fished the population 'to the ground', i.e. below its reproduction level—and it has gone extinct.

This pattern is common in many types of systems, and often leads to frustration for people who are unfamiliar with the ways systems normally behave. People often intervene in a system to eliminate a negative feedback loop they do not like, only to be surprised when a worse one takes its place. For instance, predators generally have a bad name. If you are fond of rabbits, you will try to get rid of the foxes (or hunters) that prey on them. If you succeeded, the rabbits—

as a population—would be worse off in the end because over-population and no natural elimination of the sick elements would lead to famine and epidemics. This common mistake has been actively encouraged until very recently in the field of agriculture: many wonderful chemical pesticides have been developed to kill bugs that threaten crops. Unfortunately, these chemicals also wipe out the bug's natural predators (spiders, birds, etc.). What was not anticipated is the incredible adaptation capacity of pests that soon come back in greater—mutated—numbers, and have now no predators to limit them. The result is that we then need to spray the chemicals even more often to keep the pests under control. In a way, we have become addicted to spraying chemicals. In recent years, this phenomenon has been understood by biologists who have been able to fight some pests by breeding a predator *specific to them* and using a more 'natural' way to fight the population. This policy has been particularly efficient against disease-carrying flies or mosquitoes, and shows a true understanding of the systemic nature of ecologies.

In an organizational context, trying to fight problems by eliminating what seems to hurt most is a common strategy. However, people are seldom aware that 'the road to hell is paved with good intentions'. If there is a strong underlying structure at work the 'obvious' solution is very likely to be the wrong one.

> Two head nurses founded a training company to respond to an increasing demand for hospital staff training. The two founders left their hospital positions and started doing training in various hospitals in France. They had a 'public service' attitude and offered excellent service at low prices, not being particularly interested in profits and believing quite strongly in the rightness and usefulness of their work. At the end of the first year their time was fully booked and they were turning down work. They were then joined by four of their ex-colleagues who wanted to get out of the administrative culture of hospitals. The general agreement was that trainers would only be admitted if they were up to the most exacting professional standards
>
> The quality of their work quickly gave them a reputation for people who got results with a human touch. The third year, their headcount was up to 12 trainers, and more were recruited throughout the year to cope with exploding demand. Their rule about who they would use as trainers was felt as constraining since they could not service the demand: it was limiting their growth. Therefore, they decided to lower their standard, and soon enough the headcount rocketed to about 50 trainers.
>
> The early joiners were feeling increasingly disgruntled. They did

not like the 'new' instructors. Most of them were not up to the standards of the early joiners. In a word, they were not 'good enough'. Some of the 'old hands' started leaving and finding jobs with competing companies, others would start up on their own. The founders were at a loss. Ultimately they believed that the quality of their training was solely due to their choice of trainers. They could intuitively feel that the overall quality was suffering but contracts had to be serviced, They did not realize that by getting rid of a 'constraining factor' they had caused most of their troubles.

'Limits to growth': an S-curve process

The 'limits to growth' structure combines both a positive and a negative loop. This growth process can be represented by the influence diagram (Figure 5.25).

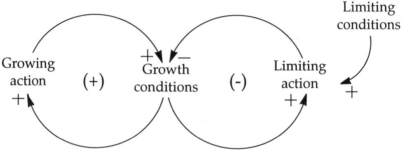

Figure 5.25 Limits to growth

This structure is the most likely to happen in real situations. As the saying goes 'when you light a candle, you cast a shadow'. In a certain context, conditions are such that a certain action is required. This action will modify the entire context—not only the conditions that triggered it. In this entire context, if our action reinforces its conditions, these conditions are also likely to trigger a balancing action. In this case, the growing process feeds on itself to produce a period of growth or expansion (positive loop). Then growth begins to slow down and eventually comes to halt (balancing effect on the negative loop).

Niche strategy

A company will grow by exploiting a new niche, and will limit itself as the niche becomes saturated. Many managers staunchly believe that a firm ensures its survival—its continued ability to secure resources—through finding a niche, a localized monopoly of some product–market combination away from the evils of competition. *Thus strategy is seen as a search for a profitable niche* (Figure 5.26).

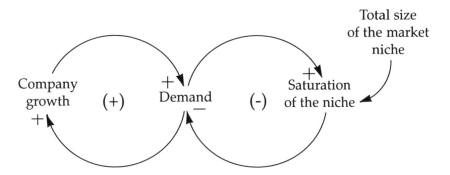

Figure 5.26 The limits of niche strategy

Unfortunately, the same managers will also believe that the success of a firm is reflected by its growth rate. However, growth cannot be sustained on one single niche, hence most of management's headaches. As an example, let us consider morale in a growing company. In the rapid growth phase, morale is high. Talented junior members are highly motivated. But as the firm grows larger, its growth slows down. This means fewer promotion opportunities, more in-fighting among junior members and an overall decline in morale.

For instance, a few years ago, I witnessed the birth of a new company. Its founder had developed a new idea for atmospheric captors in the meteorological field, which turned out to be very useful to monitor air turbulence at ground level on aircraft landing fields. In the first years, growth was phenomenal, demand was very strong, profits were high, and the firm attracted a number of brilliant young engineers who made incredibly rapid careers in terms of promotion and salary rises. One of the processes of that company was that you got promoted every two or three years as the company was growing at a tremendous rate and desperately needed bright people. Morale was high and the place was full of fun and excitement. Unfortunately, such a promotion 'policy', since people tend to stay long in a successful job, meant a structural imbalance between the number of joiners and the numbers of managers: there were never enough joiners and already too many managers, the firm was getting extremely top heavy, and the only way to maintain this unbalance was by a sustained exponential growth.

After a few years, the firm filled its market niche. There just was no more room to expand. Yet, it continued to promote its people on the same basis of fast promotion. The ratio of managers to staff grew to the point where the creator opted for a 'purge' and by different means 'got rid' of several managers and reinstated an acceptable situation. The consequences to morale were disastrous, and to his dismay, six months later, he found himself in exactly the same situation. As the process works, the system will generate its own behaviour regardless of where you want it to go. In this case, the firm's founder had to resign himself to 'bureaucratize' his firm by creating more hierarchical levels, longer promotion times and accept paying the price in terms of falling motivation and rapidly losing his most gifted young people, while realizing that, for a while, he needed to stabilize his growth even if that meant using fewer high-profile staff.

Winners learn, losers do not

The difference between winners and losers is that winners learn and losers do not. All very well, but learning is actually difficult, and progress tends to follow an S-curve. For instance, imagine picking up a tennis racket. At first, you will make tremendous progress, be able to hold the racket, hit the ball, etc. Then you will start hitting the real difficulties of tennis, and your progress will slow down. It is at this stage that many people get extremely frustrated and give up. Well, the same applies to any change process (Figure 5.27).

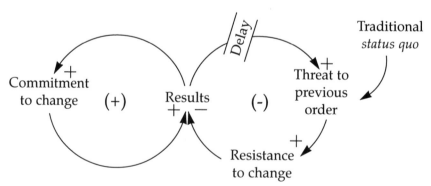

Figure 5.27 Resistance to change often appears as change is gaining momentum

The difficulty organizations face in changing, or adapting, lies in the limiting factor of their prevalent mental models. Those representations of the world constitute their norms, their reference of 'normal behaviour'. And whereas change may be indulged at first, after a while the balancing loop can kick into action and slow down the change process. It can even reverse it. At this point the more you push for commitment, the stronger the resistance. As Machiavelli wrote:

> There is nothing more difficult to take in hand, more powerless to conduct or more uncertain in its success than to take the lead in the introduction of a new order of things because the innovators have for enemies all those who have done well under the old conditions and lukewarm defenders in those who may do well under the new.

Many 'quality programmes' have failed after having an initial positive effect. Often, a big noise is being made about quality— seminar, speeches, training, etc.,—but the structures of goals, objectives and remuneration are unchanged. At first, the people believe what they are told and are very committed to the idea of quality (Figure 5.28).

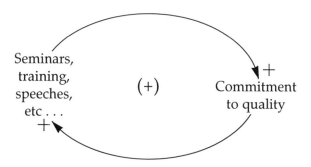

Figure 5.28 The idea of quality

But very soon, the old imperatives will reassert themselves. People will experience 'the hard way' that it is actually against their self-interest to apply the quality concepts they have been taught; and this for several reasons. First, quality involves reflection—which means putting some time aside to think about processes and how to improve them. Although this will lead to a more efficient workplace overall, the pressures of the job are still there and managers will frown at this idea of 'taking time off to think'. Secondly, it is rare that improvement does not go against some rule or other (which explains why things are the way they are in the first place). Without proper empowerment, and true commitment from management, these rules will be very difficult to 'short cut'. The resulting feeling from the staff is one of discouragement, disbelief and cynicism—and they go back to doing what they have always done (Figure 5.29). Until, a couple of years later, the new MD decides that something must be done about quality, and now instigates a 'culture change' programme (Figure 5.30).

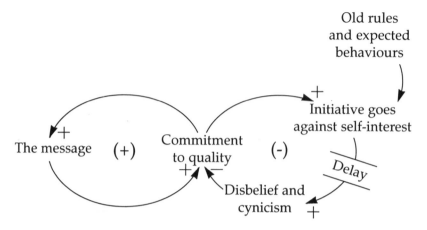

Figure 5.29 Difficulties of applying quality concepts

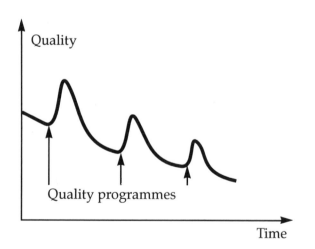

Figure 5.30 The short-term effect of quality programmes not followed up by structural changes

A major bank went through a total quality management (TQM) programme, and sent most of its staff to attend quality seminars in groups of 20. When the bank tellers came back to their branches, they found that they were to use their initiative in satisfying customers provided that they would follow the bank's rules and procedures. They were actively discouraged from questioning the procedures, even if they felt that they got in the way of good service. Management's attitude was 'there is a good reason for this, none of your business, just do your job'. The enthusiasm built up by the programme was soon replaced by widespread cynicism and low morale.

Most systems are composed of a multiplicity of coupled positive and negative loops

Complex systems are made up of a multiplicity of coupled positive and negative loops. This tends to make them 'counter-intuitive'. Imagine you are driving a truck with several trailers attached to it. As long as you are going forward, there is no problem, the trailers follow. If you take a turn, however, you will have to be very careful because of the inertia of the trailers. This describes the *delay* in complex systems. A sharp turn at the front can cause considerable grief down the line.

Furthermore, imagine that you are now trying to reverse the truck and the trailers. Even with one trailer this is a difficult operation because if you are going right in reverse, the orientation of the wheels will tend to send the trailer left and so forth. Complex systems are counter-intuitive because they can be compared with a truck and several trailers trying to reverse. One does not know what consequences to expect.

However, in practice, we can identify some of the driving loops. Unless we model the whole system on a computer (which can be very difficult and time consuming) we will never know for sure—my mind cannot cope with much more than four loops at a time. Yet, by drawing out what we feel are the strongest loops in the system, we can get a 'feel' for the dynamics of it, and for where the pressure points are.

Nothing can grow forever

Most of the time, positive loops will be balanced because the action itself uses up a scarce resource somewhere in the system. A limit to growth constraint can be a resource constraint, or an internal or external response to growth. Nothing grows forever, although we often expect it to.

Most organizations tend to develop a strong 'business as usual' attitude. The feeling is that if we keep on doing what we have always done well, things will turn out OK in the end. This attitude leads to a particular form of decision making: incrementalism. The idea is to stick as closely as possible to the previous decision made on the matter and to modify it only slightly. Unfortunately, in some cases, this policy only makes matters worse.

Fundamentally, any growing quantity can approach its carrying capacity in four generic ways:

- As long as its limits are far away, or are growing faster than itself, any quantity can grow exponentially without interruption (Figure 5.31).

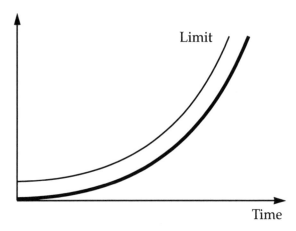

Figure 5.31 Exponential growth

- If the limit is near but our growing entity receives accurate, prompt signals about where it is with respects to its limits, and if it chooses to respond to this information quickly and accurately by slowing down its growth, then it will level off smoothly into an S-curve and maintain the balance (Figure 5.32).
- If it does not respond accurately to the signals—or if the signal is delayed or distorted—it can overshoot its limit. If the limiting factor is renewable, then the system will oscillate around the limit (Figure 5.33).
- Some overshoots, however, are irreversible. Nothing can bring back an extinct species, nothing can revive a *bankrupt* company. 'Bankrupt' actually means exactly that: stretched past banking limits. If signals or responses are delayed and the limits are erodible, the system will irreversibly degrade itself (Figure 5.34).

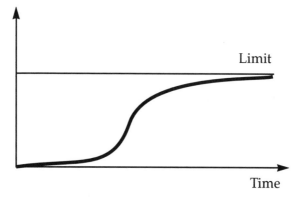

Figure 5.32 Growth and stabilization

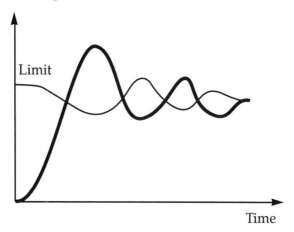

Figure 5.33 Overshoot and rapid response

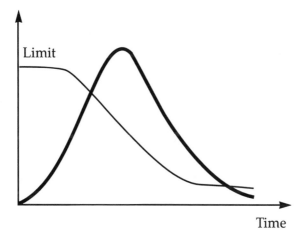

Figure 5.34 Overshoot and collapse

The only way a complex system can finally collapse is if it permanently depletes a scarce resource necessary to its survival (for instance, if we used up *all* the oil resources, our civilization as built on the car would collapse). Fortunately, most complex systems are quite resourceful in dealing with what threatens them most. This understanding of systems behaviour is what led the economist Malthus[10] to formulate his famous theory: the end of civilization as we know it is near because population grows faster than our food production capacity. As it happens, and fortunately for us, we have been able up to now to increase our food production capacity faster than populations.

Things, however, are getting strained. Neo-Malthusianism pops up here and there when we consider very real problems such as pollution and our natural resources depletion. The question is still the same 'have we grown beyond our limit?' Population explosion, for instance, is one of the most bandied around doomsday scenarios. However, some social experiences of places such as Madagascar show that exploding population can lead to a 'developed' industry, which then seems to auto-regulate its population as most Western countries do. The truth is that there is much work to be done in the systems field to move away from superstition towards learning. Often we find that the real question is about *efficiency* rather than *quantity*. The real question is not 'how many humans can this planet sustain?' but 'what standard of living is acceptable for the bulk of the population?' Similarly, the problem is not to 'ban technology, stop mass production to solve our waste, pollution and resources problems' but 'how do we best use existing technology to make mass production efficient?' Without a systems framework and understanding, most of these issues seem intractable and become a matter of opinion or dogma rather than rational thought.

Overview

Central to the systems thinking framework is the notion that structure drives behaviour. Systems thinking identifies four basic behaviours: exponential growth—or decay; goal-seeking behaviour; cyclical behaviour; and S-curve growth. These behaviours are characterized by the following underlying structures: positive, or reinforcing, loops; negative, or balancing loops; negative loops with delays, and finally a combination of positive and negative loops called 'limits to growth'. These behaviours are explained in detail with a number of business applications and lead us to operational action points.

Notes

1. For a clear description of the mathematical basis of GST see the appendix in L. von Bertalanffy; *General Systems Theory, Foundations, Development, Applications*, 1968, George Braziller.
2. I have not included chaotic behaviour in this book for several reasons. Firstly, it is not clear yet what the scope of Chaos theory is in practice. Secondly, some systems dynamicists argue that chaos is a mathematical effect that is not proven 'for real' in the real world. It is my personal opinion that chaotic behaviour can be observed in some organizational incidents—but only in very specific and marginal situations. I have therefore chosen not to include chaos in the present discussion. For a very interesting application of Chaos theory to organizations see: R. Stacey, *The Chaos Frontier*, 1991, Reed International Books.
3. In M. Keough and A. Doman, 'The CEO as organisation designer', *The McKinsey Quarterly*, 1992, No. 2.
4. This is a special case of the 'bandwagon effect' which was accepted in the post-war years as an economic behaviour. For more on the bandwagon effect see: H. Leibenstein, 'Bandwagon, snob and veblen effects in the theory of consumer's demand', *Quarterly Journal of Economics*, 1950, No. 64, 183–207.
5. See J. Bleibtreu, *The Parable of the Beast*, 1968, Victor Gollancz.
6. G. Bateson, *Mind and Nature*, 1979, Wildwood House.
7. C. Karass, *The Negotiating Game*, 1970, Thomas Y. Cowell.
8. In P. Senge, *The Fifth Discipline: The Art and Practice of the Learning Organisation*, 1990, Bantam Doubleday.
9. See J. Forrester; *Collected Papers of Jay Forrester*, 1975, MIT Press.
10. T. Malthus, *First Essay on Population*, 1798, reprinted MacMillan 1926, 1966: 'The constant effort towards population, which is found to act even in the most vicious societies, increases the number of people before the means of subsistence are increased. The food therefore which before supported seven millions, must now be divided among seven millions and a half or eight millions. The poor consequently live much worse, and many of them be reduced to severe distress.' And so on.

Chapter 6

Leverage

One of the profound insights of systems thinking is to show how structure will influence behaviour. Structure influencing behaviour means that different individuals placed in the same structural situation will tend to behave in similar ways. Of course that is not exactly correct, because we all know that individuals are different from one another. However, by and large we can make this assumption—which is mostly corroborated by our experience in organizations. Consequently, in order to influence how people behave, we shall endeavour to modify the structure in which they operate. Seeing the structure and understanding how it works is the first step. Modifying it is the hard bit. To quote Peter Senge:

> The bottom line of Systems Thinking is leverage—seeing where actions and changes in structures can lead to significant, enduring improvements. Often, leverage follows the principle of economy of means: where the best results come not from large-scale efforts but from small well-focused actions.[1]

Jay Forrester has spent 40 years studying systems interactions in companies.[2] His view is that, most of the time:

- Whatever the problem is (falling market share, unstable inventory, inadequate quality control) it is nearly always traceable to the way the company *does* things.
- Often one small change, in one or a few simple policies, will solve the problem easily and completely.
- The high-leverage policy point is usually far removed in time and place from where the problem appears. It is seldom the subject of much attention or discussion, and even when it is identified, no one will believe it is related to the problem.
- If it happens that someone has indeed identified and questioned the high-leverage policy, that person has almost always decided to push the lever in the wrong direction, thereby intensifying the problem.

As it happens, we tend to focus on symptoms where the stress is the greatest. The area where the pain is the greatest tends to end up at the top of our 'to do' list. In fact, we try to fix or improve the symptoms. However, as we repeatedly experience (with great frustration) such efforts only make matters better in the short run, at best, and worse in the long run. In fact, instead of fixing the symptom, we might well try to reach the root cause, and find the fundamental solution.

Taiichi Ohno is the mind behind the Toyota Production System. A fundamental tenet of the system[3] he put in place was to make sure that every employee had the means of:

1 Identifying a problem as early as possible in the process.
2 Finding the root cause of the problem.

Massaki Imai, a Japanese total quality consultant reports:

> In the factory, problem solvers are told to ask 'why' not once, but five times. Often the first answer to the problem is not the root cause. Asking why several times will dig out several causes, one of which is usually the root cause. Taiichi Ohno, the former Toyota motor vice president, once gave the following example of finding the real cause of a machine stoppage.

> Question 1: Why did the machine stop? Because the fuse blew due to an overload.
> Question 2: Why was there an overload? Because the bearing lubrication was inadequate.
> Question 3: Why was the lubrication inadequate? Because the lubrication pump was not functioning right.
> Question 4: Why wasn't the lubricating pump working right? Because the pump axle was worn out.
> Question 5: Why was it worn out? Because sludge got in.

> By repeating 'why' five times, it was possible to identify the real cause and hence the real solution: attaching a strainer to the lubricating pump. If the workers had not got through such repetitive questions, they might have settled with an intermediate counter-measure, such as replacing the fuse.[4]

Furthermore, by using a symptomatic solution, we usually trigger a rather unpleasant side-effect, that will make it even harder to implement the fundamental solution. By spraying pesticide over crops, farmers use a symptomatic solution that kills all the 'good' insects in the process. Because pests do not have any natural predators, we are now addicted to using pesticides. This structure can be represented by the influence diagram (Figure 6.1).[1]

There are three clues to the presence of this type of structure: First, there is a *problem that gradually gets worse over the long term*—

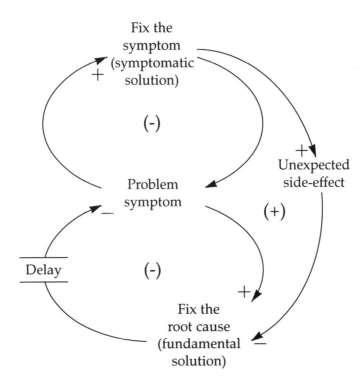

Figure 6.1 The squeaky wheel approach: oil it where it squeaks loudest

although every so often it seems to get better for a while. Second, the *overall health of the system gradually worsens*. Third, there is a *growing feeling of helplessness*. People start out feeling euphoric—they have solved their problem—but end up feeling as if they are victims.

In fact, this structure describes the mechanism of addiction. When we use alcohol, tobacco or pills to relieve stress, we do not address the fundamental cause of our stress (too much work, bad relationship, etc.) but simply relieve the pain of it. Unfortunately, we become more and more reliant on the 'fix' without ever solving the problem that *keeps coming back*.

> An advertising company decided to outsource its quantitative studies to outside consultants. The company consequently stopped recruiting anyone with quantitative skills, and soon found itself absolutely incapable of handling any data-driven problem. Thus, the company had to rely heavily on consultants who were able to keep their prices high for trivial tasks because no one in the agency could do them. The consultant would come in, do the analysis, acquire the learning, produce a report, get paid and leave. The agency was addicted to consultancy.

Some fundamental solutions can be very difficult to implement, mostly because of the delays involved. In that case, the symptomatic solution might be necessary in the short term—but keeping the fundamental solution in mind! Furthermore, fundamental solutions can be hard to find because they tend to be away from the symptom, both in place and in time. Typically, a problem in a process is caused further back. As it ripples through the process, the problem escalates, and what starts poor gets worse (Figure 6.2).

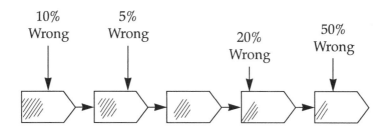

Figure 6.2 Errors ripple through a process

In this case, the information ripples backwards through the delivery process to create havoc at the production end of the chain. Megadrug is a well-established pharmaceutical company. It has a wide range of products distributed over the world. One of its newer products is a homeopathic drug, Homeocold, to stop colds from evolving into real nasties. Megadrug supplies retail shops from various regional centres. In March last year, at the end of the winter season, orders from retailers unexpectedly started to pour in. The regional centres serviced the shops and ordered the plant for more Homeocold. Unfortunately, the production line at the plant had already been switched to another 'summer' drug, and it would take at least a week before Homeocold could be produced again.

As the regional centres stocked out, the retailers were frantically ordering more Homeocold. Megadrug's management team realized then that a media-triggered 'poisonous drugs' scare had resulted in millions of customers switching from traditional drugs to homeopathic alternatives. After a crisis meeting, they decided to switch production back to Homeocold in order to service the increasing backlog with the regional centres. Even at full production for one month, they would not be able to catch up; but one thing Megadrug could not afford was a reputation for unreliability with retailers and customers.

Weeks later, the regional centres finally started servicing the retailers again. To their surprise, they were greeted with relief at first, but, as the weeks went by, with embarrassment. Retailers started cancelling orders, faster and faster. The regional centres, watching piles of Homeocold accumulate in their warehouses, started screaming at the plant to stop producing and shipping Homeocold.

The plant director was pulling out his last remaining strands of hair. He was now asked to stop producing Homeocold and to catch up with the summer production that had got way behind schedule. The management team was too busy blaming each other to notice that Homeocold was still selling steadily more than usual and decided to delay production next year to get rid of the stocks at the regional centres. By increasing expectations and ignoring the obvious delays involved, the company faced a 'Forrester effect'. Minor market changes at the end of the supply chain accumulated to cause great havoc with production planning. Once again not reacting would have been safer than over-reacting.

Identifying leverage points

There is no hard and fast way to identify leverage points. Mao Tse-tung once said:

> The most important problem does not lie in understanding the laws of the objective world and thus being able to explain it, but in applying the knowledge of these laws actively to change the world.

Indeed, in most real-life systems leverage is not obvious to most of the actors involved. They do not see the structures underlying their actions; they are caught by their own agendas, objectives and constraints. Systems thinking can give some broad rules that will prove useful in a 'leverage-hunt'. These guidelines tend to appear obvious and sound like truisms when we read them, but the real difficulty lies in their practical application.

Most of the time, systems thinking will be useful for avoiding making disastrous mistakes rather than finding the cleverest, optimal policy. Its view is one of long-term survival, not short-term gain. However, we still often expect *a solution to be near a symptom, a long-term gain to start off with a short-term gain, or a winning strategy to bring instant gratification for all players.* We know that complex systems do not behave like that, that things will get worse before they get better, but something within us insists that they should.

Don't panic!

People react too much, and too quickly. Most organizational actions have very long time-spans compared to individual time-spans. In our experience it is likely to take 14 to 18 months between the moment a signal is received to when action is actually taken. The time-span of a human being is closer to the day and the week, sometimes the month. It makes it very difficult to appreciate organizational change and not to over-correct when we take action.

Organizational problems are particularly difficult to handle because we cannot easily distinguish the effects of our actions. The situation is too complex, causes are multiple, and the time-spans are very long. In these circumstances, people will tend to rely on ideology more than on experimentation: I do this, because I believe this is the right thing to do. These actions can lead to very surprising and unexpected results, hardly the desired ones. However, there is sometimes no other way to act than to rely on a rule of thumb. Yet, at the same time, we will jump to conclusions too fast, and react too violently. This immediate emotional response (our emotions do not understand time-spans) is not always appropriate and can make matters worse.

> The managing director of a successful components manufacturer has a very simple human resources policy: hire winners and fire losers. In practice, being seen as a loser means termination. Unfortunately, this MD is on a rather short fuse and any failure is seen as a dismissible offence. Some problems have nothing to do with the individual, but, you will still get fired for them.
> Consequently, the clever executives learn many astute ways of covering up failures, even when they are legitimate. As a result, problems tend to appear too late, when nothing can be done about them. The MD then has to step in, fix them as he can, and look for a scapegoat, and he keeps complaining that he spends his time fire fighting rather than dealing with fundamental strategic issues.

As Forrester is fond of saying: 'Don't just do something. Stand there!'

Anticipations

Everyone knows that what goes up must come down again, at some point. Yet our anticipations tend to be way off. All negative feedback creates oscillations—some large, some small. Nevertheless, people have trouble dealing with cycles. If the economy has been going well for the past four years, nearly everyone will be bullish. We simply project our experience into the future and expect *more of the same*.

However, the longer the boom, the more likely the recession. The first signs of slowing down should be taken very seriously. Similarly, everyone is gloomiest at the bottom of a recession, just when rapid growth is more likely. We tend to have short memories, and to project the recent past on a straight line (Figure 6.3).

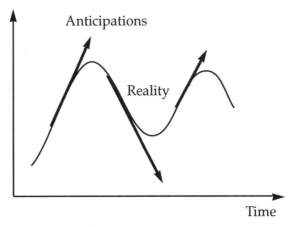

Figure 6.3 Our anticipations are projections

Such anticipations lend themselves to very poor strategic decision making, particularly when they are linked with the assumption that resources are unlimited. By putting trends into a longer historical context (seeing patterns, not events) we can raise the right questions about their future.

> A famous real-estate development firm launched a prestigious project at the peak of the 1980s boom. As the project was finally completed, the firm went bankrupt because the market had crashed and the buyers disappeared. What happened? In their anticipations, the developers made two fundamental systemic mistakes. First, they anticipated that the boom would go on for at least as long as it would take them to develop their new complex (a minimum of five years). Secondly, their estimation of the market did not take into account the projects of other developers which would be finished sooner, and hence attract a huge percentage of demand. They anticipated the office market as though they were the only player in it!

There is no blame, no enemy, no 'out there'

If we see that systems cause their own behaviour, then we can get rid of blame. There is no point rejecting causes outside our boundaries; it is the 'they' syndrome: *their* fault, *their* problem, *they* did it ... And it is *us* or *them*! *This* us and them attitude is one of the most pervasive aspects of human societies. Our sense of identity tends to derive from the exclusions of 'others'. Interestingly, no matter how small the circle gets, we always find someone to exclude, to reaffirm our own sense of identity. This can lead to very strange situations, where two groups can be satisfied to be obviously losing—as long as the other loses at least as much as we do.

Masaaki Imai tells the following story:

> Thirty years ago, Kaoru Ishikawa encountered this problem head-on while employed as a consultant to Nippon Steel. In one instance, Ishikawa was investigating some surface scratches found on certain steel sheets. When he suggested to the engineer in charge of that particular process that his team review the problems together with the engineers of the following process, the engineer replied, 'Do you mean to tell us that we should go examine the problems with our *enemies*?' To this, Ishikawa replied, 'You must not think of them as your enemies. You must think of the next process as your customer. You should visit your customer every day to make sure he is satisfied with your product.' However the engineer insisted 'How could I do such a thing? If I show up in their workshop they'll think I've come to spy on them!' This incident gave Ishikawa the inspiration for his now famous phrase: 'The next process is the customer.'[5]

Managers will sometimes express the feeling that they are limited because 'the organization' or its staff is not fast or efficient enough to enable them to do their job. In fact, the manager is part of a system. If staff members are already overworked, there will be delays in their delivery. Put the pressure on and botch jobs will appear, which will need reworking, increasing the backlog. But for some managers, it is not their problem (Figure 6.4).

As long as managers do not include their staff in their system, they are unaware of the vicious circle created by their staff's backlog: the situation will only get worse. Tensions will rise, tempers flare and their own ability will be crippled. By expanding their system to include the staff, managers will be able to *see* the problem, and thence, solve it (Figure 6.5).

What is the next limiting factor?

Growth itself modifies the factors of growth. Growth in a complex

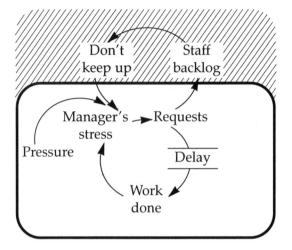

Figure 6.4 Manager's perception

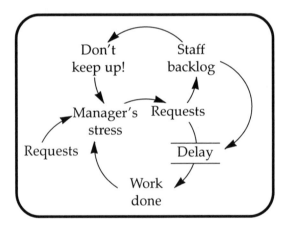

Figure 6.5 Manager's system

system can require a multitude of inputs, and we tend naturally to focus on the greatest demand. However, at any one time, there is only one important factor: the most limiting one. This does not necessarily mean the largest one. As we said, bread will not rise without yeast, no matter how much flour you add on to it. The most limiting factor is the one that is going to block any further progress.

The IT department of an international FMCG company was about to launch a major system development project. This system would ultimately compile data from all the countries where the company was trading. The project steering committee spent months developing a plan in painstaking detail, and yet a month before the starting date they still hadn't decided who the project manager would be.

Experience shows that the completion of IT projects depends more on the personal qualities of the project leader rather than on detailed planning and in that case it was the most limiting factor. However, the planning was the largest task to be done before the project would start, and consequently was given priority.

Leverage can be found by looking for the next potential limiting factor rather than monitoring the abundant ones, before the balancing loop becomes too strong. Other than a strictly speaking limiting factor, your system may have an implicit goal. Many of the successful corporate culture studies revealed the implicit goals in organizations. The point is that the most limiting factor is not necessarily the 'biggest' factor, nor the most quantifiable. 'Goodwill', for instance, can be a strong limiting factor to the commercial success of a company, yet, in most people's mind it will only be secondary to the marketing budget, or the promotions plan, and so forth. To quote Donella Meadows:

> This concept is childishly simple and widely ignored. American economists have claimed that energy cannot be an important factor of production because it accounts for less than 10% of the GNP. (Yeast accounts for much less than 10% of the bread—that doesn't make it unimportant.)[6]

Change the rules, change the game

Complex systems are inherently goal seeking. Understanding the structure of a system—identifying the positive and negative feedback loops—is important because if we just 'kick' a system, our action will have a short-lived effect after which it will come back to its stable position. Any change—no matter how big—which does not modify the structure of the feedback loops will have only a temporary effect. At the same time, any change, no matter how small which affects the links between parts of the system is going to change the long-term behaviour of the system.

An IT consultant in Australia relates how he was hired by a firm to build a system that would help the company keep track of their workers' punctuality and presence. The company operated a punch clock system which generated enough data to keep five persons busy collating it. They thought that a computer system would help them with better analysis and fewer overheads.

After considering the problem, the consultant suggested getting rid of the clock system altogether and proposed workers' salaries instead. Rather than keep track of the hours actually punched in by the workers, they would be paid a salary at the end of the month and have the responsibility of keeping the machines running. After initial uproar and indignation, management finally accepted the proposition.

As it happens, productivity went up in the plant. Previously, a worker would switch the machine off, go to his locker and change and then punch the clock at the given time. The next shift arrived, punched the clock, went to the lockers and changed and then had to re-start the machine before any work could be done. By suppressing the punch clocks, workers would adjust between themselves so that one could pick up from the other without having to switch the machine off. Furthermore, if they had extraordinary delays, workers would 'cover' for one another, so that the machines were kept producing full time. Consequently, the department collating the information from the clocks could be shut down and no IT system was needed.

Do not fight positive feedback; support negative feedback instead

Most of the time, when what we are trying to do does not work, our immediate reaction is to do the same, but harder. It is what we call the 'bigger hammer' theory. If it does not work, try harder. Unfortunately, the harder you push, the harder the system pushes back. Many of our best intentioned actions suddenly boomerang. Do not try harder, but try more intelligently!

This is how dogma justifies its failures: if a policy does not show the promised results, it is always possible to argue that the policy was not pursued relentlessly enough to make a difference. From a systems point of view, when growth slows down, we are likely to find a balancing loop in action. This loop can be created either by a physical limit, or by an implicit goal. Instead of pushing for growth it will be far more effective to expand the limit, or modify the goal, in order to weaken the negative feedback loop.

The case is similar if you are trying to thwart the growth of something unpleasant. Trying to limit the elements of the positive feedback loop will, at best, slow the growth but not stop it; and this at a considerable expense. Fundamentally, the harder you push, the harder the system will push back; according to the laws of action and reaction. Creating and nurturing a negative feedback loop can be far more effective in the long run. In medicine, prevention is more effective than cure. If we fight an epidemic, we need to cure people as fast as they contract the illness. On the other hand, if we manage to isolate the illness, we can then cure it without the risk of exponential explosion.

> A very practical application of this principle is when we reach a deadlock with someone, and instead of yelling louder and pounding on the table we ask *'what could I do that would make it easier for you to do this?'* I have found that whenever I ask something out of the ordinary of someone, this little phrase works wonders. Very often the person will then explain what her or his most constraining factor is and we will find a way around it. On the other hand, if I make a scene, the person feels aggrieved and is much less likely to give in to my demands.

Shorten the delays

Most traditional planning processes involve long delays. From orders to deliveries, the time-span in certain cases can reach several months. For instance, from the customer, the order must go through the retailer's inventory, then to the distributor's warehouse, then to the factory's warehouse; through the factory, out of the production line, the product goes back to the warehouse; then to the distributor, then to the retailer before reaching the customer (Figure 6.6).

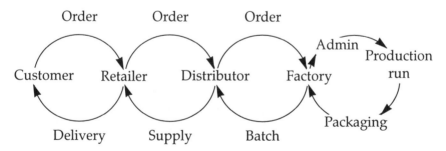

Figure 6.6 Supply chain

If we map out our physical and information processes we can find the critical delays and improve the efficiency of the whole system. George Stalk Jr considers that time is the next source of competitive advantage and that companies in the 1990s have to learn to compete on a time basis[7]—in other words, by shortening delays in everything they do.

He describes the case of Atlas Door, a ten-year-old US company that has grown at a 15 per cent annual rate in a sluggish industry. Atlas is doing remarkably well on the basis of its time mastery. First it shortened production cycle-time by building a just-in-time factory. Then the front-end was attacked and shortened by streamlining and automating its entire order-entry, engineering, pricing and scheduling processes. Thirdly, Atlas tightly controlled logistics so that it only shipped fully completed orders to construction sites. By these several actions, Atlas managed to reduce its lead time to a few weeks in an industry where the accepted norm is almost four months. In a decade, the company replaced the leading door suppliers in 80 per cent of the distributors in the country.

Project management

In fact, project management—as a general management approach—was first highlighted by Forrester when he pointed out the delays involved in traditional corporate management of long-term tasks. Functional subdivisions in the organization across the product flow necessarily involve long delays. At each step, the decision-making process must go back to the central 'head' before gravitating down to where the work is actually done. This not only creates too many hand-offs from one department to another—which create delays and inefficiency—but also a bottleneck at top management level (Figure 6.7).

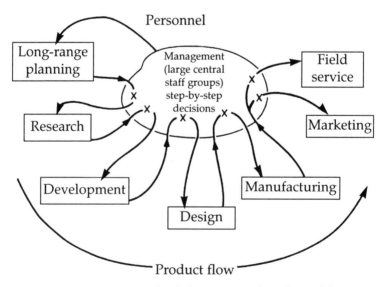

Figure 6.7 Functional subdivision involves long delays

Forrester suggested a different form of organization: 'project organization'. In this organization, the product flow is not interrupted by constant back and forth between top management and the functional departments. Work flows both faster and more consistently from one step of the project to the other (Figure 6.8).

> In the project organization, top management takes a view that is longer than the individual project. Control is exercised through the initial approval of project objectives and scope, by approval of the budget, and by approval of the person to take charge of the project. Top management exercises continuing control by reviewing changes in the planned budget and scope, by observing progress, and by the passive action of permitting the project director to continue unless matters take such a serious turn that the director must be replaced or the project terminated.

> The project organization makes it easier to act on the necessary early preparatory steps that anticipate the problems of the later product stages such as manufacturing or marketing. To be successful, the project organization must of course have the project leadership able to exercise the required skills in each of the functional activities. The perpetual expansion of the leadership is one of the longer-term processes involving the management structure.[8]

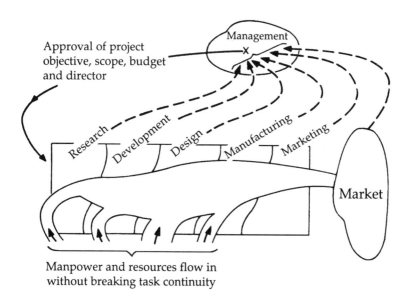

Figure 6.8 Project organization

Overview

If one sees that structure drives behaviour, one can see that some small structural modification can have huge impacts, where large diffuse actions might not significantly affect outcomes. The art of systems thinking lies in identifying these 'leverage points'. Although complex systems tend to be counter-intuitive and leverage points difficult to find, practice has highlighted some areas where the systems thinker might try to look first. These areas include delays, anticipations, the next limiting factor, and focusing on the negative feedback loop rather than fighting the positive feedback. We must realise that unless we modify the structure itself we will not change the system's behaviour in the long term. In other words, change the rules, change the game!

Notes

1. P. Senge, *The Fifth Discipline: The Art and Practice of the Learning Organisation*, 1990, Bantam Doubleday.
2. J. Forrester, *Collected Papers of Jay Forrester*, 1975, MIT Press.
3. See J. Womack, D. Jones, D. Roos, *The Machine that Changed the World*, 1990, Macmillan.
4. M. Imai, *Kaizan*, 1986, The Kaizen Institute, McGraw-Hill.
5. Ibid.
6. D. Meadows, 'Whole earth models and systems', *The CoEvolution Quarterly*, 1980.
7. G. Stalk and D. Hout, *Competing Against Time*, 1990, Macmillan.
8. J. Forrester, *Industrial Dynamics*, 1961, MIT Press.

Chapter 7

Steps to a systems thinking analysis

Conducting a systems analysis is not always easy. People tend to resist abandoning their traditional assumptions about the world, and have to buy into the systems framework. Furthermore, because of its very nature, it can be difficult to place boundaries over some systems problems. This is a proposed approach—not a prescription. Practice shows that most cases can be handled in this framework, but to each his own.

1 *What is the symptom?* How does the problem or opportunity appear? How does it look, sound, feel? What is the history of the symptom? Can we map it?
2 *Who are the players?* Who is involved in this? What are their logic bubbles? Can we draw a motivation map? And an action matrix?
3 *What is the growth engine?* What is the positive loop representing the driving forces? What is in people's minds?
4 *What are the main limiting factors?* What are the main balancing loops? Is there a rare resource which we are ignoring?
5 *Expand the frame.* What is the next link? The previous one? Is there a loop outside our perception?
6 *Operational action points.* Where can I find the leverage? What practical action can we take?
7 *Communicate!* Having discovered the magical answer is not enough. Others must now buy into it if anything is to happen.

What is the symptom?

If purpose is a main element of mental models, context is also of prime importance. Most of us tend to think either too generally or too specifically. It is difficult to find a way to go from the general to the specific and vice versa. Classical thinking tends to go from the general to the specific, the problem or question is defined in the vaguest, most general terms such as 'what can be done about the fruit industry?' and then the question is narrowed down until we can provide some answers. This narrowing down process tends to be somewhat arbitrary and to follow well defined and used brain tracks. The production expert will argue about how to produce more and better fruits, the finance wizard will find a way to have a greater return on fruit investment, the philosopher will try to define what exactly we mean by 'fruit', and so forth.

Personal biases are very strong in this narrowing down process because of the obvious temptation of going down the road where we might have some answers ready. The alternative is mind-boggling. How can we be expected to take into account all the elements of the problem? What is more, most problems are divergent rather than convergent. A 'convergent' problem is a nice problem in which the more we study the question, the closer we get to a solution. Unfortunately, most problems in life are 'divergent', which means that the more you study the problems, the more new and difficult issues appear, and the further away from any solution you get. Studying a divergent problem is like digging in the sand, the more you dig, the more sand you find. With these types of problems, the tendency is to focus on the little bit we know how to handle, regardless of how relevant it is.

For this reason, it is far better to define precisely the problem in the first place, and then to expand it if we find that the way we defined is too narrow. By defining the problem precisely we can use a relevance criterion rather than just a 'we'll treat this bit because we know how to deal with it' approach. Rather than thinking about 'what can be done about the fruit industry?', we can think in terms of 'what could happen to the fruit industry that would affect us'.

Keep it relevant!

Relevance is always the best guide, it does not discriminate between internal or external, short term or long term, measurable or intuitive. It singles out what we feel, a priori, will have an impact, and we can then proceed to be proved right or wrong, to expand our field of investigation or to be satisfied with a rough answer. Asking oneself 'what if I choose to move to another country?' is a question without an answer, or more accurately too many answers. If we rephrase the same problem in terms of 'what events might lead me to move to another country?' then we can start defining the problem in earnest, and from then on, start solving it. Using relevance as a criterion is also a way of uncovering problems or decisions that might not appear immediately in the operational routine of things. Some questions about the future of your particular industry, consequences of choices or easily dismissible accidents that yet tend to occur increasingly often will not be 'problems' as such, they will not be full-blown crises. But they might become so. Relevance is a way to spot and monitor what is really important in the general area we are looking at. Most decisions come down to a 'Yes' or a 'No', yet their consequences can be vast and engage our entire future. Rather than confronting the 'entire future', it is much more practical to identify the decision and trace its consequences.

It is precisely because systems thinking is holistic by nature that we need to be rigorous in the defining of the question. In all the various techniques outlined in this book, we shall always insist on a specific starting point which progressively expands rather than on a general problem to be narrowed down to the specific. Systems thinking is focused on including as many relevant elements in one specific problem rather than on high level, vague, general answers. It is about confronting the situation with its context rather than trying to assess the 'whole context'.

> In practice, we find it useful to start with a chart representing the behaviour of what we are looking at over time. In some cases, the chart can span a great number of years—rather than the current tendency to consider only the last three years. This chart does not need to be precise, we are only trying to get a 'feel' for the variable's behaviour.

Who are the players?

Organizations are people working with other people regarding people and for people. They do not exist in themselves, they do not walk down the streets. None the less, we tend to forget this obvious statement and treat organizations and systems as if they were 'real' beings. In the last analysis, humans are mammals, highly clever ones maybe, but still mammals, and as such they share some characteristics with the rest of the group. Their main interest is about their relationship with their group as a whole and with specific individuals. When we talk to our neighbour about the weather, we are hardly sharing any information since both are experiencing the same weather, but we are very actively communicating 'mammal' messages such as a willingness to talk to each other, social position symbols, etc.

However, our main preoccupation in life, i.e. how we see others and how others see us is curiously absent from most of our 'rational' thinking—to the extent that we tend to make decisions which involve others without consulting them. In a few cases this is deliberate, but in most cases it simply does not occur to us. As a rule, in our crowded environments, each time we make a move we are likely to affect someone else, yet, for some complex and obscure psychological reasons, we tend to ignore the fact or cover it up under the term 'self-interest' which is not too good, but after all our common lot. In many ways we deal with our difficulty to control what others will think of us by denying that the others are persons, we confine them to an organizational role: my boss, my client, my secretary, my engineer. Then, when people get angry, frightened, depressed, hostile, frustrated, or offended we are often surprised. 'What's biting him?', 'What's the matter with her?' and throw a tantrum about 'professional behaviour'. Other people have egos that are easily threatened; they see the world from their particular vantage point and often confuse their perceptions with reality. More often than not, they interpret what you say according to their fears rather than understanding what you meant. Misunderstandings will reinforce prejudices, which will in turn create more misunderstandings, and so forth. The name of the game becomes scoring points, confirming negative attitudes and allocating blame. This is costly to everyone involved, and somehow, once started, nobody can figure out what went wrong.

> Problems or decisions do not exist in isolation, as abstracted entities. Rational thought will often treat them as such and probably find a rational solution. Typically, all hell breaks loose at the implementation stage and no one understands why. More often than not, people who are affected by the decision have not been involved in the process and when faced with a sudden action which engages them, they react violently.
>
> The following quote attributed to a General Motors executive by Mintzberg and Waters illustrates this point: 'We use an iterative process to make a series of tentative decisions on the way we think the market will go. As we get more data we modify these continuously. It is often difficult to say who decided something and when—or even who originated the decision ... I frequently do not know when a decision is made in General Motors. I do not remember being in a committee meeting when things came to a vote. Usually someone will simply summarize a developing position. Everyone either nods or states his particular terms of consensus.'[1]

Who will be affected by this?

The first step in finding out what a problem really is about is to identify who is involved, directly or indirectly. Who are the 'stakeholders'. In any social or organizational situation, any decision will affect a number of persons or groups. It is of the utmost importance to answer the basic question 'who will be affected by this decision?', because in most situations the reaction of these people is critical to the success of an action.

Some of these groups will be internal, some external. What is important is to identify them as 'actors', i.e. groups of people who will tend to share common attitudes and can have an impact on the problem by choosing a certain course of action. In a company the main groups typically are:

- **Internally**
  - Accounting
  - Finance
  - Management
  - Support staff
  - Personnel
  - Production
  - Research and development
- **Externally**
  - Customers
  - Suppliers
  - Competitors
  - Influencers
  - Regulators
  - Unions

These different actors will have specific concerns and interests and will be able to act upon the system in various ways. It is important to study them in more detail and identify what characteristics of an actor are relevant to a systemic analysis.

The motivation map

The motivation map is a visual tool to draw out the main concerns of each of the actors. Interests usually tend to define the problem, and we seek to identify the main interests of each actor rather than to go deep into the details of the motivations themselves. In doing so, we get a holistic picture of the situation: where the main tensions will be found; where there is common ground, and so forth. We tend to assume that because of conflicting positions all interests are incompatible. In most cases, however, an examination of the interests will reveal more interests that are shared or compatible than interests opposed. In a selling situation, the buyer and the seller have opposing interests concerning the price, yet they have a stronger mutual interest in making the transaction, maintaining a good relationship, and so forth. The aim of the motivation map is to identify these outstanding interests.

The simplest technique to identify interests is to put yourself in other people's shoes. When trying to understand an actor's concerns, simple role playing will usually uncover his or her basic interests. Most actors are explicit about their interests, they express them through their position at issue. By examining each position and asking yourself 'why?' the interests are often obvious. As a rule, the most powerful interests are basic human needs. These needs, strangely enough, also apply to organizations, or more accurately, to those who run them. Fundamental human needs include:

- Security
- Economic well-being
- A sense of belonging
- A sense of achievement
- Recognition
- Control over one's life
- Control over one's territory.

As basic as they seem, they are easily overlooked. In many cases, we tend to think that the only issue is the money because it is usually where real conflict surfaces and it is easily measurable. Other basic interests tend to be involved in the process and are expressed in monetary terms. By treating them independently, the picture which emerges is often more coherent.

To draw the motivation map, we place the interests on a sheet of paper. Each actor is represented by a bubble with his or her interests written in it (Figure 7.1). Common interests are found in the intersection between the bubbles. The easiest way we have found of building such a map is by writing all the interests on Post-it stickers and then moving them around the actors' names until they draw a satisfactory representation.

MANAGEMENT

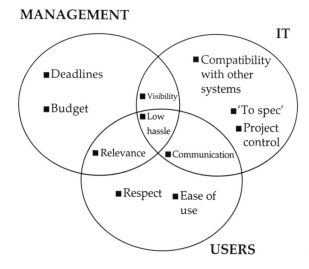

Figure 7.1 A motivation map

I believe what I see, I see what I look at

Interests alone do not define action. Actors act according to their perception of personal interests as opposed to what they perceive their current situation to be. In systems terms, actors monitor certain conditions which will control their action. Often these conditions are gaps between their goals and the present state of the system; but the very way these goals are measured influences what action is to be taken. Whether one measures the commercial success of a company by its sales turnover, profits or share price, the picture one gets is likely to be widely different.

Each concern needs to be expressed in terms of how the actor evaluates it. What does he or she measure, monitor, act upon? In some cases, this might be disconnected from the concern. Monitoring sales turnover as a measure of commercial success to evaluate an interest of economic well-being could be quite perverse: the company could be selling at cost and therefore losing money on each item sold. Yet the concerned actor would not take action on the basis of his or her perception that sales are growing, therefore there is nothing to worry about. Identifying the conditions associated with each concern enables us to draw the action map.

Identifying the key actors immediately changes our outlook on any problem. Classical analysis tends to study problems as abstractions, disconnected from the reality of people with wants, fears, hopes and power. Everyone has power: the power to do, or not to do; to agree or disagree; to change; to learn. People are part of the solution and should be treated as such. Classical analysis has three blind spots in this respect. Firstly, it does not take all the actors into account, and might miss a critical one; secondly, it confuses persons with roles, people with their position. This confusion is largely sustained by the people themselves. If you ask someone 'what are you?', they are likely to answer 'I am a carpenter', or a tinker, tailor, soldier, beggarman, thief. And yet, are they merely that? As far as most organizations are concerned, they are. Thirdly, as a consequence, we get upset by problems but since we do not talk to the others involved, our actions tend to reinforce the problems.

Where is the invisible man?

Most problems flare up as conflicts. Most decisions aggravate someone somewhere. People argue, crisis is declared and steps are taken. It is unpleasant for everyone involved, but what can you do? Surprisingly, from a systems point of view, the people who are locked into a conflict over an issue are often not those responsible for the situation. The man who is involved in an accident because someone ignored a red light and disappeared will have to cope with the damage and the aggravation; as will staff members who are at each other's throats because their supervisor believes in 'divide and conquer'; workers in conflict with their management over a problem caused further up the production chain. The cause is elsewhere. We focus on the symptom, and as normal human beings, argue with the closest person around—not very productive, but it makes one feel entirely justified. In many cases, there may well be a hidden actor that creates the situation but does not appear directly in the conflict. Often this actor is himself unaware that he is creating chaos and destruction somewhere else.

Because, psychologically, we have such a propensity to jump to conclusions, it is good practice to take the time to think about who is involved in any situation other than those obviously there. Once the actors have been identified, then we must concentrate on the interest underlying their positions. As with children fighting over an orange, we might discover that one wants it for orange juice and the other to use the peel in a cake. Interests are the key to the purpose of people's action, but this purpose can express itself in obscure ways, through misrepresented facts and twisted logic. As with driving, the danger is not so much in the car you see, but in the one you do not see.

A manager was asked by his director to study the consequences of a new possible course of action. The deeper he went into it, the more complex he found it and came to the conclusion that although this course of action was rich in opportunities, it would demand much more research to ensure its success. Busy as ever, he forgot to mention it early to his director who one day summoned him to his office and asked him for a simple answer 'should we do it or not?' When the manager started to explain the problems involved, the director threw a fit and said that he expected results, not excuses. The manager stuck to his guns and said that he could not honestly give a realistic opinion as yet. In the end, it surfaced, that the director himself had been summoned by the board to give his opinion, and that all he needed was something to tell the board, not a decision. As it turned out, the real issue was that the director himself was under fire. Not being aware of that, the manager thought that the value of his work was being challenged and took an unnecessary self-righteous stance that was very damaging to the relationship. The director would have been happy with a preliminary report he could present to the board that said: 'we need more research!'

What is the growth engine?

The driving positive loop is usually easy to draw out, because it is the one people focus on most naturally. Whenever you look at a situation, most people can describe 'what makes it work'. Planning is mainly done on the basis of creating a positive loop and hoping that it will work out in practice.

> It is often quite easy to start from one particular action, and look for the condition it needs to reinforce it. In practice people tend to think in sequential steps, first this, then that. As Lewis Caroll wrote:
>
> 'Cheshire-Puss,' she [Alice] began, 'Would you tell me, please, which way I ought to go from here?'
>
> 'That depends a good deal on where you want to get to,' said the Cat.'[2]

Like Alice in Wonderland, managers have had it drummed into them by consultants, business schools and trainers that they must identify objectives before doing anything. Ted Levitt wrote one of the most influential articles in the marketing world stressing the importance of beginning with ultimate objectives rather than operational goals.[3] Many companies have adopted 'management by objectives' as the way to do things. Most meetings in my experience do start with the obligatory 'what are we trying to achieve?' In fact, most of the decision making and planning seems to work on the basis of:

(1) identifying where we are; (2) defining where we want to be; and (3) deciding how to get there.

This is a very sound procedure in itself and a necessary one. Decisions and action plans therefore tend to be built as a series of steps that will bring us from point A to point B. This can be laid out in considerable detail using planning techniques such as PERT charts and so forth. At each stage, the plan outlines what needs to be *done*.

> A department in a large administration wanted to upgrade its word-processing capability. It: (1) conducted a survey of word-processing packages; (2) chose one on predefined criteria; (3) implemented its decision by buying the software and training the staff; and (4) evaluated the results. This plan sounds fairly 'natural' yet the implementation failed—twice. The first time the software was never even bought, and the second time the staff continued to use the old package until it was forcibly removed from the machines. However, all decision makers in the room at the time of the decision were in complete agreement with the decision (the need to increase word-processing capability by changing packages) and with the plan.

What are the main limiting factors?

The main limiting factors can be identified by looking for the scarce resource used up by actions that appear in the positive loop. Many of these limiting factors can be 'softer' organizational issues. Although the softer issues of organizations are seldom expressed explicitly, managers tend to be very aware of them—even if not consciously. A manager will have an instant feel for 'what can and cannot be done'. He or she will also be faced with the structure's implications concerning how time and resources can be prioritized. Therefore, many difficult decisions will be accepted as logical on the face of it but are likely to create intense psychological pressure on the individual managers as they go about implementing the decision.

Each manager will be confronted by a set of 'limiting factors' which are the practical expression of the structure of the organization. These factors are usually of a very practical nature—and at the same time hard to express. The main difficulty I have encountered in managing change is a situation where we ask a person (1) to take time off his or her regular job to perform another role and (2) to modify his or her behaviour. Both requests can be accepted by the person as perfectly logical and legitimate since they are issued by the manager. They can also appear impossible. The individual may already be overworked—so how can he or she be expected to take on yet another role—and the new behaviour asked for is not only difficult to acquire, but also probably goes against the accepted way of behaving in the peer group. At that stage, we usually demand that he or she accepts on trust that the new behaviour will enable better performance so that there will be more time for the new role, from which he or she will gain increased status which will legitimate the new behaviour, and so forth. None the less we must accept that these requests will create anxiety because, in a way, they go 'against' the behaviour induced by the structure.

Limiting factors are relative and diffuse

The difficulty we encounter if we try to highlight these limiting factors is that they are very relative to the organization and the individual. Since they are an expression of the structure, they tend to differ from one organizational role or status to another. People have different objectives, responsibilities and constraints—and therefore their limiting factors will be different, linked to their specific position in the structure.

Limiting factors also depend on the individual since different personalities will have preferred problems: situations they can tackle better than others. Consequently, two managers in the same organizational position can be confronted by very different limiting factors. This difficulty is accrued by the fact that we are dealing with *perceived* difficulty. The fact that these limiting factors are not formally addressed in the decision-making argumentation means that the individuals have no reference to inform them whether the difficulty they anticipate is realistic or not. In some cases, managers will tend to ascribe intentions from their own fear and increase the perceived difficulty of the situation.

Confronted with one of these limiting factors, the individual manager can either: (1) go ahead regardless of his fears; (2) do nothing and wait to see what happens with the other parts of the process; or (3) substitute his or her task with something he or she feels will fulfil the same role while being easier to achieve. In combination, these behaviours can inflict a considerable drift on a project and still not be noticeable. Because of their very nature, these behaviours tend to be diffuse: they will express themselves on different, apparently unrelated issues—and as a result will not give a consistent picture of what is happening. Faced with this kind of drift no one can ascertain one cause of drift and fix it. We will then look for wider explanations such as the 'non-adaptive culture' of the company—explanations which can cover a multitude of sins. The prescription in such cases tends to be a 'culture change programme'; and again, unless such an activity-based programme tackles the fundamental structural issues of time and resources allocation, recognition and reward and so forth, it is likely to make no more than a superficial impact. The main difficulty with such actions is that they cannot be achieved locally. Suddenly a small implementation problem can mushroom into a company-wide culture problem and will be dropped with a sigh (*how can we expect to get anything done in this type of culture?*)

Expand the frame

Remember to think in terms of processes, not events. There is usually a number of people or departments involved in a single process, and they might not be evident (like the invisible man). It can be useful to explore upstream and downstream of the process chain to find what *really* drives this system.

> For instance, in managing a cross-functional project a project manager lived with the fear that people would point-blank refuse to do their part of the work because it conflicted with their own functional priorities. As it turned out, they never did, and the job got done. However, he was having a different problem throughout the period. The overall manager of the department would call the project manager in regularly to tell him to be 'more careful' and to make 'fewer waves' and so forth. As a change agent, he dismissed such comments as part of the job, until they became rather strong and personal—and to the point that he almost stopped everything altogether several times, not knowing what was causing such uproar because the people around him maintained they were quite happy with how the project was developing.
>
> It was only after the project was (successfully) completed that it was possible to understand what was happening. There was no real problem within the project team itself (no more than expected) but the managers of each of the persons involved resented that the cross-functional project was taking precedence over some of their functional priorities. Although the project manager had authority to run this project which was considered important overall, several functional managers would complain to the project manager's departmental manager about their lack of resources, work not getting done, the project manager's 'negative

influence' on their staff, etc. The departmental manager responsible for the project would then come to the project manager about vague—but strong—complaints that he could not relate to the work itself. Whereas the project manager expected resistance from the work itself, nothing happened, but the reaction came from a different angle—and almost stopped him from finishing the project.

Structural action points

That is where the magic touch of systems thinking lies. Sometimes the leverage point is obvious: by exposing the limiting factors, people find out right away what needs to be done, and how to do it. At other times, leverage cannot be found right away—so try, and try again. As a rule it is important to keep in mind that structure drives behaviour, and that unless the very structure is modified, little is likely to happen.

Michael Goodman, Vice-President of Innovation Associates, proposes the following interventions on a causal loop diagram:[4]

- Add a link.
- Break a link.
- Shorten a delay.
- Make a goal explicit.
- Slow down a growth process; relieve a limiting process.

Disprove your mental models

Find opportunities to test your mental model of the situation. When we work in any situation, we have an, a priori, understanding of it. The tendency is often to take any excuse to reinforce this perception rather than to modify it; communication should be used to disprove our mental models rather than reinforce them, we need to focus actively on the aspects which 'do not quite fit' and to understand why they do not fit in order to build a more accurate map of reality. In this frame of mind, most of 'communicating' is about listening actively and acknowledging what is being said. By actively we mean in an active mode, proposing an idea and then focusing on the response, acknowledging the difference, and building a common representation.

> Most people tend to glorify thinking, rather than simply treating it as a tool—particularly intelligent or clever people. The strategic planning department of an oil company was given the opportunity

to try out in practice one of their boldest propositions. For multiple reasons, they let the opportunity go by. Years later, reflecting on this experience, one of the team believes that they were actually scared of testing their hypothesis for fear of being wrong. So they kept the idea as basically valid—but left it on the shelf.

When Galileo built one of the first telescopes, contemporary theologians refused to *look* into it for fear that seeing the stars would contradict their understanding of the universe. In our thinking, we tend to think in terms of either/or. In fact, no one mental model represents reality accurately. True intuition comes from accepting apparently contradictory representations of the same situation and finding the 'point of view' that will make them fit. Failing to do so often means discarding vital opportunities.

Norman Dixon relates how the tank—the most significant military technological advance of its time—was consistently dismissed by British high command. Since its invention in 1908, the tank was used once, with reluctance, as a cavalry squadron during the First World War. After a disaster at Cambrai in 1917 where no one had anticipated the speed at which tanks could penetrate into enemy territory and therefore no follow-up was ready and the offensive was lost, the tank was dismissed from further consideration. An officer tried to show that new strategies were needed with new technology, even wrote a book explaining these new strategies, but was consistently dismissed by the establishment and ultimately expelled from the armed forces. However, this new thinking was not lost on everybody. To quote Norman Dixon: 'Unfortunately Liddell Hart's entry was not entirely lost to view. Along with other products of his pen, it was enthusiastically studied by Hitler's Panzer General, Guderian, and became required reading of the German General Staff.'[5] In this instance, the British military were unwilling to disprove their mental models in search of a better solution.

Communicate

This is not always the easiest part of the analysis. Nevertheless the buy-in of stakeholders is vital if we want to see anything happening. It is worth mentioning that it is often easier to sell an experiment rather than sell change. Furthermore, when you sell an experiment, you can involve the stakeholders in both refining your analysis of the situation and defining the criteria for success—so as to establish the ground rules beforehand.

Talk to them, not to yourself

Similarly it is important to talk to the other actors, not simply to ourselves. It is easy to forget that dialogue and debate are two very distinct things. Our culture tends to understand communication as debate, one party should be proved right or wrong by the strength of its arguments. What is good for the courtroom is not very helpful to avoid getting to the courtroom at all. To establish a real sharing of perspectives, ideas and concerns one must not only listen but talk actively to the other actors in a language that they will understand and not be offended by. In the corporate world, there is a strong temptation to employ a barrage of techno-cultural terms that show how important and clever one is but do not actually make sense to anyone else. Production talks in production terms, finance in finance terms, and so forth. It might even be necessary to adopt a common language to evolve a common representation. The first step, however, is to make the effort to understand the other's language and to purge our own of our hermetic technicalities. Most things under the sky can be expressed simply and clearly.

Talk about the whole system

Do not focus exclusively on your own problems or concerns. Try to give the 'whole picture' and show how all the actors are involved in the system and have an interest in getting it to work. In doing so, acknowledge the particular concerns of each individual actor. Then build on the common ground. From the motivation map, it is often possible to show how individual action leads to perverse effects which go against mutual interest. The motivation map explicitly shows the common ground and how each actor is involved in the system. As a rule, if talking to a specific actor, speak about yourself and not about him or her. Do not judge, condemn or evaluate, but simply express how you feel. A statement about how one feels is very difficult to challenge. It conveys the information without provoking a defensive reaction which will reject it.

Be specific

In order to avoid confusion, it is desirable to be specific. Instances and examples are more effective than general descriptions even though they are usually more difficult to express. To describe a system accurately is rather difficult. We find that the easiest way to go about it is: (1) to describe the mechanism in abstract terms; (2) to choose a specific, commonplace example; and (3) to apply this to the situation at hand. Being specific also entails asking specific, pointed questions. The consensus on interviewing tends to be for the 'open questions' such as asking 'why?', 'where?', 'how?', and so on. In practice, open questions tend to be far too open, and the other person just gets lost, or starts answering a question in terms of his or her own immediate preoccupation. When exploring an issue, rather than going from the general and narrowing down to the specific, here again it is often easier to start from a specific point and explore around it. Systems thinking is a pragmatic, down to earth approach. In its communication also, the emphasis should be on specific, practical descriptions of behaviours, motivations and actions.

Overview

This chapter presents a method we have developed and tried out in practice in various systems thinking workshops. It is by no way prescriptive, but can help systems thinking novices to explore an issue thoroughly before they can develop a method of their own. The seven steps include expressing the symptom in terms of dynamic behaviour; identifying the key players; identifying the growth engine and its obvious limiting factors; expanding the frame of analysis; identifying possible action points; and finally communicating the results of the analysis to other parties involved.

Notes

1. H. Mintzberg and J. Waters, 'Does decision get in the way', *Organisation Studies*, 1990, Vol. 11, no. 1.
2. L. Caroll, *Alice in Wonderland.*
3. T. Levitt, 'Marketing myopia', *Harvard Business Review*, July–August 1960, no. 38.
4. M. Goodman, 'The do's and don'ts of systems thinking on the job', *The Systems Thinker*, August 1922, Pegasus Communications.
5. N. Dixon, *On the Psychology of Military Incompetence*, 1976, Jonathan Cape.

Chapter 8

A systems representation of organizations

Organizations: we spend our working days dealing with them, and our non-working hours trying to escape them. Can systems thinking help us make more sense of these monstrosities and see the persons behind the roles. The key element of organizations as we know them is accountability. Work is organized under the principle that some people (managers) are accountable to others (their bosses) for what further people do (staff). With this simple structure, we can go on to build rather grand pyramids in which everyone is more or less accountable to somebody. Accountability tends to be defined by two parameters: the goals or objectives each individual is accountable for; and the processes defined by the organization through which this individual will achieve those objectives.

In this vision of organizations the relationship between processes and goals is extremely important. In truth, in complex organizations, several types of 'processes' are at work. These include organizational rules as stipulated in company policy, more implicit rules such as 'cultural' behaviours reinforced by peer pressure in a given group, and finally technical operating procedures. From a systems thinking point of view, we shall take a very broad sense of the word 'processes' that would, in a way, cover *the way things are done* in an organization.

In real life, people sometimes make processes up as they go along, but only in new, uncharted areas, and the legitimacy to 'create' processes is often limited to a few very senior people in the organization. Precisely because of the size and the complexity of our organizations, there is usually very little room for individual freedom. A member is supposed to perform by achieving his goals through activating the prescribed process to the best of his ability. In systems thinking terms, a member of an organization is part of the feedback loop (Figure 8.1): the gap between the prescribed objective and the actual result will determine more use of the process until the gap progressively disappears.

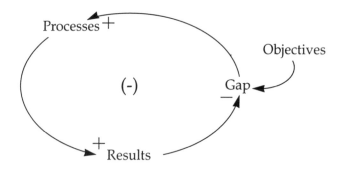

Figure 8.1 The organizational loop

Objectives and processes also account for most of the organization's troubles when they either have to be modified because of a change in the context or, more often, interfere with each other in a way which is detrimental to the overall outcome. This is what we commonly term 'organizational inertia', or 'momentum'. In fact it is just normal systemic behaviour where the interaction between the various goals and processes leads to their natural conclusions. In these cases, the system will have to face *decisions* (Figure 8.2).

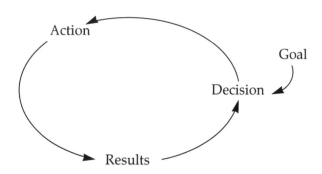

Figure 8.2 Decisions

Or, to establish the appropriate *corrections* (Figure 8.3).

Many organizational policies work on the basis of '*incrementalism*': to quote Janis on the subject:

> If verbalised, the rule runs like this: 'Stick closely to the last decision bearing on the issue and make only slight changes to take care of the most urgent aspects of the problem currently at hand.'[1]

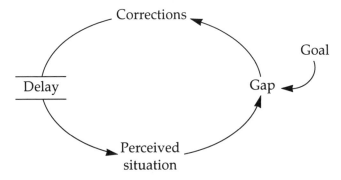

Figure 8.3 Corrections

The 'systems' of the organization

If we refer to the 'organizational control loop' and focus on all its applications within one complex system, we will find underlying codes which determine how the organization works (Figure 8.4).

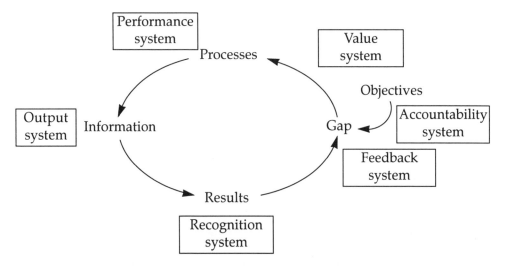

Figure 8.4 The systems underlying the organizational loop

The people confronted with this organizational loop will look for guidance in six fundamental structuring 'systems' or consistent sets of organizational codes:

- Accountability system
- Recognition system
- Performance system

- Output system
- Feedback system
- Value system

In fact, the great danger is that often these different systems are not *aligned*. They are inconsistent with each other, and therefore pull in different directions which ultimately results in organizational chaos (or gravity). A value system is a very good integrator. It can be used to 'align' each of the other systems with regard to fundamental goals and desired behaviours of the organization. At each level, we must draw out *what people are asked to do*. Fundamentally, management is about asking people to *do* things. People are often glad to oblige, but feel frequently puzzled by what is expected from them in terms both of tasks and of responsibility. The accountability system determines both the scope of their jobs and the tasks they have to accomplish.

Accountability system

In a well run organization, the accountability system should be consistent with the organigram of the firm. Unfortunately, many organizations seem to accumulate levels of hierarchy without reference to levels of authority.

Eliott Jacques defines the level of work of a manager by the time-span of his longest task:[2] the longest of the maximum-target-completion-times of tasks in the role. He argues that people consider their 'real' manager to be at least one time-span category removed from their own.

- Everyone in a role below three-months time-span feels the occupant of the first role above three-months time-span to be the real manager.
- Between three-months and one-year time-span the occupant of the first role above the one-year time-span is felt to be the real manager.
- Between one-year and two-year time-span the occupant of the first role above the two-year time-span is felt to be the real manager.
- Between two-year and five-year time-span the occupant of the first role above the five-year time-span is felt to be the real manager.
- Between five-year and ten-year time-span the occupant of the first role above the ten-year time-span is felt to be the real manager.

The recognition system

The recognition system is, literally, *what recognition do people get for good performance*. From a purely *behaviourist* point of view, people can be controlled by the recognition (or lack of recognition) they get for adopting 'good behaviour'. Without promoting the 'carrot and stick approach' of management, we can recognize that most people are clever enough at least to pretend to act in a way that will bring recognition.

Recognition can then be taken in the larger sense of reward, and expressed according to any of these dimensions. Different people will value different interests differently: some will be sensitive to security (*now you've got tenure!*); to money (*How about a nice bonus?*); to group belonging (*You're one of the lads now*); to acclaim (*Hey, you're a hero, well done!*); or to increased responsibility and control over one's life (*You've earned the right to do this job on your own*).

> Recognition can at times be garbled and lead to double bind: messages of the type 'do what I say but don't do what I do'. In one organization, junior members were constantly encouraged to 'speak their mind' and be innovative. However, the people who were promoted to manager were invariably those who were 'safe', 'reliable' people. When a junior would go to voice complaints about this state of affairs it was explained that if he or she was not progressing fast enough it was because there was something lacking in his or her behaviour. After a few such experiences, juniors soon learned to keep their mouths shut, or left the firm.

Recognition can often account for 'implicit goals'—situations in which operators are asked to do one thing, but rewarded for another. This situation, common in many organizations, creates great stress on the operators—but is also usually resolved in the sense of best interest. As a result many 'strategic initiatives' may fail if they are not accompanied by a modification of the recognition system. It is hard to ask someone to 'go and do better quality' and at the same time reward him or her at a strict 'per hour on the chain' rate.

The performance system

Close to corporate culture, the performance system is *what good behaviour looks like*. Again, this may conflict with what is expected— and what is rewarded. Nevertheless, establishing what good behaviour looks like is of critical importance in most groups. We often find that different groups value different types of behaviour. Many anthropologists have shown how different cultures value opposite behaviours.

For instance, such traditional cultures as the Navajo American Indian value respect and discipline. Trying to teach Navajo children to be more competitive to fit into the American school curriculum is almost impossible. Their parents will not favour them for the behaviour they are expected to have in classes.

> Anthropologist Claude Lévi-Strauss tells the story of missionaries arriving in the Amazon Jungle and being killed on sight by the natives.[3] Finally someone discovered that baring one's teeth was read as a sign of aggression by certain tribes, who would massacre the missionaries welcoming them with broad smiles at first contact.

> Most people in organizations need to prioritize the demands made on their time. Sometimes, managers can get frustrated because something they have explicitly asked for does not get done. Very often, we find that action is not part of the behaviour expected from people in that company, and therefore is put at the bottom of the 'To Do' list. A manager may have great difficulties sending staff on training courses. Although he or she will know that staff development is critical to the success of the firm, the staff will always find something more immediate, or more important to do than going on training courses.

The output system

How results are measured will influence greatly what good behaviour looks like—or which part of behaviour 'shows'. If non-smoking is part of good behaviour, but there is no way of finding out whether someone smokes, one relies purely on the goodwill of the smoker—and his or her capacity not to smoke!

> In several firms I have been faced with the issue of trust. At the beginning of workshops, people would assert that they need more trust between departments. Yet, when asked what 'trust' looked like, sounded like and felt like, they could not describe it. As a consequence they often realize that they do not know whether people are trustworthy or not because they have no way of evaluating it—hence an atmosphere of general distrust.

Charting the results over time is a fundamental component of continuous improvement. For instance, let us suppose that we feel that our reps should make more telephone appointments before going to see customers—in order to avoid making the trip for nothing if the customer is not available. Rather than say 'you must make more phone appointments' and waiting to see a result, we can establish the graph shown in Figure 8.5.

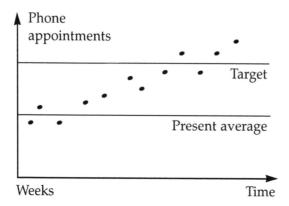

Figure 8.5 Charting results over time

A few years ago, a car manufacturer used to measure the output of its factories by their quarterly financial results. When a car was shipped to a dealer, the dealer's bank account was direct-debited for the price of the car. Because of the competitive atmosphere between the plants, it became very important to have ever improving results. Consequently, at the end of a quarter, the plant would send any old car on their premises to dealers, debit the account, and expect to see the car returned and exchanged next period. Unfortunately this mechanism snowballed, because it needed to work with larger amounts each quarter, until the whole scheme fell apart. Here, the output system is driving the behaviour of each of the actors—to catastrophic end.

The feedback system

The feedback system tells *people how well they are doing*. If this system is not properly set up, the actors are blind, and can go on pushing an action way past its usefulness into overshoot. The great problem about feedback is that it should be immediate and to the point. Regrettably, in most organizations, feedback on how someone is doing tends to take place after long delays. Think about the evaluation reviews which happen twice a year—if you are lucky.

Immediate feedback is a main component of job satisfaction, yet it is too often ignored by managers who seem to assume that their staff can read their minds. Feedback should be:

- Immediate
- Specific (about observed behaviour, not perceived attitudes)
- Constructive (proposing an alternative)

Charles Keating tells the following story:[4] Peter is a telephone repair man in a large phone company. The company specializes in installing and maintaining phone networks for businesses. Peter is highly respected by his boss and many of the other repairmen for his expertise and friendliness. During the last four years, the company set up an evaluation programme for the service it gives to customers. This consists of a written evaluation given to the customer with the bill. It asks questions such as whether the customer is satisfied with repairs done and with the service given, and what he or she would like to see improved.

About 20 per cent of the customers take the trouble to complete and return the evaluation, but Peter has never seen one. Evaluations are sent to the manager, not to Peter. Peter does not feel very concerned about his job because he never has any feedback on it apart from a yearly review and a dressing down when something goes wrong—which is a bit strange because, after all, Peter is the person the customer is going to meet; not his manager.

The noise in the system

One of the greatest difficulties with feedback is to be able to discriminate between 'noise', i.e. random variations in numbers, and meaningful differences. Very often, people focus on an event vision of things and react to feedback immediately, without putting it into context. Many arbitrary decisions result from attributing to a specific reason (or person) something that is purely caused by luck.

Author Raphael Aguayo describes quality guru W. Edwards Deming's 'Red Bead Experiment'.[5] In this exercise, eleven people are asked to role play a typical work environment: six workers, one foreman, and an administrative staff of four. This is a modern factory, using the latest state-of-the-art technology. The factory makes white beads, but sometimes it turns out red beads, which are defects. The customers only pay for the white beads.

The production equipment includes two plastic rectangular pans. One is larger than the other and contains 4000 beads—3200 whites and 800 reds. A rectangular paddle with 50 holes is part of the equipment. The paddle is dipped into the beads and, when raised should be filled with beads. Preferably white beads. A quality goal of only two red beads per paddle has been established by a very concerned management.

After a full year the number of defects per worker is as follows:

Names	Q1	Q2	Q3	Q4	Total	Rank
Ken	8	10	12	9	39	5
Barbara	6	4	11	7	28	1
Lenny	11	11	11	8	41	6
Noburu	8	11	8	11	38	4
Cathy	15	5	12	4	36	3
Steve	5	9	9	10	33	2
Total	53	50	63	49		

From looking at this table of 'quarterly results' one could easily draw conclusions. Barbara is definitely our star worker, but she has to watch herself. She slipped at Q3 but got back on track at Q4. Steve is definitely underperforming, and going from bad to worse. Cathy, on the other hand, started low but shows an impressive learning curve—she reaches best performance, and so forth.

Of course, from the way the experiment is set up, we know that the performance of these individuals is entirely due to chance. Yet, for the managers of this factory, these results can easily be accepted as accurate feedback: as Deming puts it: 'Management has asked me to inform you that although you improved, it was not enough. We were prepared to close the plant but someone in management has come up with a brilliant idea. This is a stunning breakthrough in management technique. We'll keep the plant open with the three best workers. All the above-average performers will stay on, while the laggards will be let go. Barbara, Cathy and Steve will stay on. As for Ken, Lenny and Noburu, we're sorry. We all like you. Please pick up your cheque as you leave.'

Similar mistakes are frequent in any 'management by objectives' environment. Unless the data are charted using statistical process control techniques, it cannot be relied upon as feedback representative of effort. In the late 1920s, Walter Shewhart has devised a set of limits that can be calculated from the data to filter out 'real' information from random noise. In the bead factory, the average number of red beads per person is nine. The upper control limit and the lower limit are respectively eighteen and one. As long as the number of red beads in each tray is between eighteen and one, any variation from person to person is due to chance.

The value system

Fundamental collective values must underlie each of these elements to give some coherence to the organization. The value system can constitute a sort of 'legal framework' to which behaviours or decisions can be referred to. To quote Bob Haas, CEO and Chairman of Levi Strauss:[6] 'A company's values—what it stands for, what its people believe in—are crucial to its competitive success. Indeed, values drive the business.'

If the values in place are not aligned with the other systems, the risk is that people will 'distance themselves' from what they are asked to do and then engage in defensive routines to 'protect' themselves against what they need to do, but do not value. Companies can resist learning if the very values in question are not examined as part of the learning process. This is what Harvard's Chris Argyris calls double-loop learning: learning by trial and error to improve one action is single-loop learning. Double-loop learning involves examining the values that underlie corporate behaviour:

> Argyris offers an example of getting organizations to engage in double-loop learning.[7] In analysing the problem of a $7 billion decentralized organization, he helped them to look past the symptomatic problems of unclear philosophy and the meaning of decentralization, the feeling that corporate staff lacked the adequate authority to deal with the line, the overlapping responsibilities that existed among staff roles, and the feeling among corporate staff that they did not have adequate contact with the CEO.
>
> The CEO asked a task force of staff and line people to address these issues, and they made recommendations for actions to correct the problems. These recommendations were accepted with minor changes. However, no one examined the assumptions, values and behaviours that underlay problems which were apparent. 'For example, the corporate staff and divisional officers (1) adhered to policies that they described as inadequate, (2) implemented staff roles that they said produced confusion, and (3) maintained inadequate authority among the relationships between policies and staff roles. These three actions, by their own diagnosis were errors. The question is why did the top officers adhere to, implement, and maintain these errors for several years?'
>
> Argyris asked the top officers why these behaviours were happening They provided several answers which totally changed the nature of the problem being addressed. It was found that 'Individuals were adhering to errors. They were taking for granted the error as well as adhering to the errors. Errors produced under these conditions are designed errors. They are not due to ignorance. The next step was to ask the individuals what led them to adhere, to maintain, and proliferate actions that they had diagnosed as errors.' The resulting discussions illustrated many

organizational defensive routines and fancy footwork. 'Asking executives with a Model I theory-in-use to become candid can be a recipe for trouble because their actions are governed by values of win, don't lose ... These errors cannot be corrected simply by designing new actions. To correct these actions, we must first alter the governing values. This means we have to learn a new-theory-in-use. This is double-loop learning.'[8]

Systems alignment

These systems need to be *aligned*—to be consistent with each other. If not, the results of the system are unlikely to be the ones we want. The stronger system will 'pull' and determine the outcome. In many organizational cases, we are faced with the effects of implicit goals as opposed to organizational objectives. These implicit goals will tend to pull the system away from what we are trying to achieve.

For instance, suppose that a company wants to engage in total quality management to regain its lost market share to a competitor. It will try to establish the following positive loop: *profits* should be reinvested into *quality* processes which would then generate customer *loyalty* and help rebuild *market share*, which would reinforce our profits and so forth (Figure 8.6).

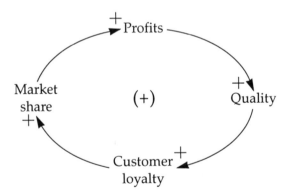

Figure 8.6 Total quality management

However, if that company's management has an implicit goal of maintaining—or increasing—market share, this positive loop will be balanced by the following negative loop: *quality* will increase the *process cost* (at least in the short term) which will affect the *profit and loss* accounts. This might affect the *share price* and push management to *cut costs* to maintain it. By cutting costs we limit quality and so on (Figure 8.7).

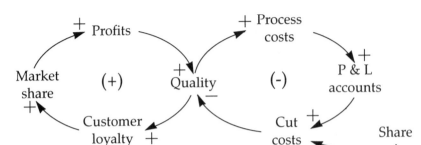

Figure 8.7 Positive and negative loops

The outcome of this process is that any quality programme is doomed to fail because it *will be cut before it has any significant effect!* The goals of management, recovering market share and maintaining share price, are not aligned.

Alignment is particularly difficult in many organizations because functional departments are built to pursue different goals. Finance is there to collate and control financial data; production is there to push products out; and marketing's role is to incite customers to buy them, and so forth. These priorities can often be conflicting. We can represent this situation as shown in Figure 8.8.

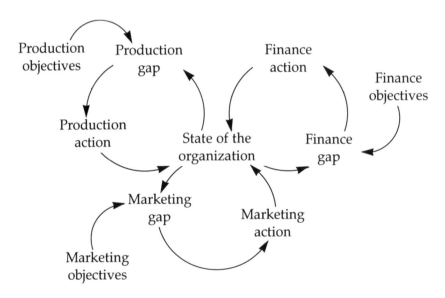

Figure 8.8 As each actor tries to achieve its objectives, the total organization falters

In the case of the organization, each actor monitors the gap between his or her objective and results and follows the prescribed processes to reduce that gap. If the gap keeps increasing as a result of someone else having the same behaviour in another part of the system, the actor will none the less continue pushing harder. The combination of all actors trying to adjust the organization to achieve all the different goals produces an outcome that is often not what everybody wants—and yet everyone is trying very hard to maintain direction because evaluation is based on actors' objectives and rules and peer pressure makes sure that they stick to the right process.

Cross-functional management

Cross-functional management is a tool to try to foster alignment within function-driven organizations. It is a management process designed to encourage and support interdepartmental communication and cooperation throughout the company. The purpose is to attain such company-wide targets as quality, cost, and delivery of products and services by optimizing the sharing of work.

> Consultant Richard Hervey relates how, apparently, CFM began at Toyota in the late 1950s and early 1960s.[9] Toyota's senior management felt the need to compete on a different basis: they would endeavour to use management effectiveness (people) as a tool for strategic advantage rather than capital-intense investments (equipment). A breakthrough in the approach was the development of a new category of company-wide functions, managed by senior line-managers—in addition to their functional responsibilities—to build a company-wide knowledge team in key areas. Over the years this system was refined within the Toyota group and has been widely applied in Japan. CFM has begun to be implemented in the US, for example at Cummins Engine, Ford, Kodak and Hewlett-Packard. The key idea is to organize around processes rather than tasks. Consequently, the hierarchies will flatten by elimination of subdivisions in workflows and other non-value added activities.

While CFM is the exception rather than the rule in Japanese companies, it is practised in firms such as Toyota, Komatsu and Pentel. In each company the implementation is slightly different, but the common theme is that there are cross-functional committees which are supported by staff of functional departments. Cross-functional committees make decisions which are then implemented by the staff of the committee, or make recommendations to senior management or the board. These recommendations are then usually approved by the board or by senior management and are implemented by the cross-functional committee staff. In this way, the functional chimneys of the organization can see where their action fits within the overall context of the company.

By understanding the workings of the fundamental control feedback loop of the organization, managers can improve their processes by having a consistent vision of their action instead of living by the 'squeaky wheel rule': apply oil where it squeaks loudest.

Overview

This chapter highlights how these systems thinking conceptual tools can be used to describe organizations and their dynamics. It focuses on the organizational control loop, and the various subsystems it implies, such as: the accountability system; the recognition system; the performance system; the output system; the feedback system; and the value system of the organization. These subsystems are essential to organizational policy design or analysis since they make up the structure of the organization: its incentives, accountabilities, systems, culture and values.

Notes

1. I. Janis, *Crucial Decisions*, 1989, Macmillan.
2. E. Jacques, *Requisite Organisation*, 1989, Carson Hall and Co.
3. C. Lévi-Strauss, *Race et histoire*, 1952, UNESCO.
4. C. Keating, *Dealing with Difficult People*, 1984, Paulist Press.
5. R. Aguayo, *Dr Deming: The Man Who Taught the Japanese about Quality*, 1990, Carol Publishing Group.
6. R. Haas, 'Values make the company', *HBR September–October 1990*.
7. C. Argyris, *Overcoming Organizational Defenses*, 1990, Simon & Schuster.
8. Ibid.
9. R. Hervey and B. Richardson, *Understanding Information, Communication, Decisions and Learning: Key Insights for Improving Automotive Management*, 1993, Sigma Associates.

Chapter 9

Change in complex systems

When we look at very large complex systems, we seek to identify certain overall properties that can help us to anticipate how they will behave in certain situations. After two centuries of social change, we can see that social systems seem to show some consistent features— yet it is still hard to pinpoint them. Very complex systems seem to have their own overall laws of behaviour. Such systems, such as very large corporations, industries or societies, are very hard to study on a loop by loop basis, but scientists, mostly from the social sciences and biology fields, come up with some consistent systemic behaviours that are worth taking into consideration.

Auto-correction

Very large and complex systems seem to develop a capacity of auto-organization. In this way, they will tend to stabilize themselves (through a multitude of control loops) around the one critical variable they are trying to maintain. Studying complex systems, Ashby shows that in a complex system, the circuits controlling the most rapidly fluctuating variables will act as balancing mechanisms to protect the stability of other variables that do not move as fast or with as much amplitude.[1] To quote philosopher Gregory Bateson:

> *Plus c'est la même chose, plus ça change.* This converse of the French aphorism seems to be the more exact description of biological or ecological systems. A constancy of some variable is maintained by changing other variables.[2]

As a result, when dealing with complex systems such as economies or very large corporations, we should not worry about the specific loops, but about their more general *capacity to change*—or, in other words, to breathe. A complex system is in danger only if this capacity starts to atrophy or if it is too reliant on one scarce resource. The upshot is usually that it will crumble into a multitude of smaller systems. We have seen this happen with very large companies bought, stripped and sold, but also with entire states, such as the former Soviet Union.

> Very large corporations tend by and large, to survive. Their very size protects them—unless their *raison d'être* disappears, such as with the coal industry. Even then, their death throes can span decades. It does not mean that these systems are stable or efficient, or permanent—simply that their very size permits enough control loop to adapt to changes in the environment in a rather haphazard fashion.

Purpose

In a way, when Rosenblueth, Wiener and Bigelow[3] proposed, in their classic paper in *Philosophy of Science*, that the self-corrective circuit and its implications provided a possible way to model the adaptive reactions of organisms, they attacked the central problem of Greek philosophy: the problem of purpose, unresolved for 2500 years. Since Plato and Aristotle, Greek philosophers were inclined to believe in what were later called *final* causes (which, by the way went on to structure most medieval Christian thought via Augustine). They believed that the pattern generated at the end of a sequence of events could be seen as the cause of the pathway followed by that sequence. This led to what they then called teleology (*telos*: the end or purpose of a sequence). This, however, did not appear to be consistent with the concept of evolution. If it appeared that a crab had claws in order to hold things, it was difficult to argue the backward causality that claws were there because they were *useful* to the crab's survival. When causal systems become circular, however, a change in any part of the circle can be regarded as a cause for change at a later time in the system: it explains adaptability to a changing environment without having to postulate prescience of the change.

> A most surprising and controversial example of such evolution appears in Lovelock's 'Gaia hypothesis'. In trying to show how the planet could have developed a consciousness, he describes the biological fact of the creation of the Earth's atmosphere. Without following the logic to its direst conclusions, we can note that this development of an oxygen atmosphere is a typical example of positive feedback loop: starts negligible and then grows on itself to saturation:
>
> 'Some of the earliest living things have left trace fossils identified

as stromatolites. These are biosedimentary structures, often laminated and shaped like cones or cauliflowers, usually composed of calcium carbonate or silica and now recognised to be products of microbial activity. Some of these are found in ancient flint-like rocks over three aeons old. Their general form suggests that they were produced by photosynthesizers, like blue-green algae of today converting sunlight to chemical potential energy.'[4] In return, oxygen provided a favourable milieu for the development of these entities, which then produced more oxygen, and so forth.

In this perspective, a system can be characterized by its purpose, which is quite different from its objectives. Its purpose is what it actually does as opposed to what it is supposed to be doing. From that point of view, the purpose of most living organisms is to survive and reproduce. An organism, therefore, will tend to adapt its behaviour in ways which it perceives will enhance its chances of survival and reproduction (its perception, admittedly, can be completely deluded).

The purpose of many old large industries is closer to maintaining jobs than to being economically efficient. This is consistent with the purposes of communities: to survive. An economic finality is much more abstract than the reality of protecting a community, a way of life—no matter how antiquated or undesirable.

Equilibrium rather than stability

There is a fundamental difference between a dynamic equilibrium and a rigid stability. One is alive, the other is dead. For instance, if we consider a pool of water where no water flows in—or out, very soon, it will become stagnant and ultimately dead. Life will at first develop in it and very rapidly choke itself to death. Fortunately, most pools are part of a system. There is an incoming flow, and an outgoing flow. The level of water in the pool can remain constant if the two flows are equal: this is a dynamic equilibrium (Figure 9.1).[5]

Such a pool will allow life to develop and to grow. It will maintain a condition of growth: *variety*. More fundamentally, it will ensure that *scarce resources* such as oxygen—necessary for life—are maintained in ample supply. Some companies or societies have tried to reach stability rather than equilibrium. Corporations try to establish everything 'once and for all'; governments will control dissension. These organizations very soon die on their feet from internal rigidification and decay.

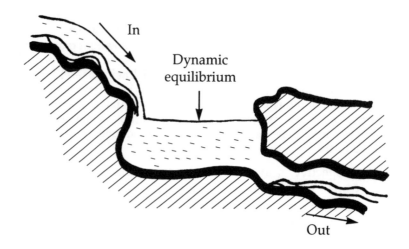

Figure 9.1 Dynamic equilibrium rather than stability

As our economies are struggling out of the present crisis, the 'no inflation!' dogma is now so strong that it is not even questioned. We consider our forefathers fools for having let 'inflation run', and so on. Yet, should we look at the question again from a systemic perspective? We need to regulate the flow of money in the economy—not to choke it. An economy can be described at a very high level as the following open system: *people's work* generates *production*, which produces *goods* (or services) which are then *consumed* and as a result create more work, and so forth (Figure 9.2).

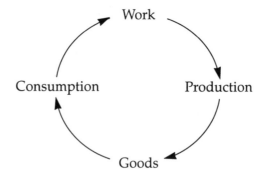

Figure 9.2 A high level view of the economy

This economic wheel is fuelled by money exchanges at each stage: *income* gives the capacity to *purchase* which creates *sales* for companies who then pay *wages* that generate more income, and so on (Figure 9.3).

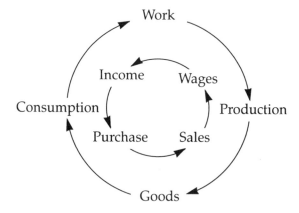

Figure 9.3 Work flow and money flow

Money is what flows through the system and maintains it alive. If the total amount of money is too tight, the exchanges slow down and that creates recession and unemployment—and a stability of prices and values of goods. However, if the amount of money in the system is too important, we will have full employment and growth—but with inflation and a devaluation of goods which leads to over-production and waste. For a healthy economy, we want to alternate both policies so that the system can regulate itself.

Bifurcations

Ashby once pointed out that to produce anything new, a system needs to contain some source of randomness. This randomness can be reinforced by positive loops that will create a divergent system. In fact, although systems have a tendency to auto-correction, sometimes some random element can grow out of pre-existent positive loops—previously held in check—and push the system towards a completely different equilibrium. This will not happen often, but is rather sudden and violent when it does.

This randomness is often created by an environmental change. A regulatory change, for instance, is likely to modify the structure of an industry. It can also simply be the trigger point of a system which cannot hold its internal pressure—and often appears in the form of revolutions. In the 1970s, researcher Ilya Prigogine[6] introduced the concept of evolution through bifurcations: a system can evolve by steps—without possibility of return to its previous state (Figure 9.4).

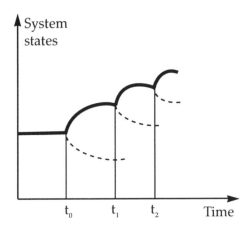

Figure 9.4 Bifurcations

These bifurcations are extremely hard to predict. Some random element is going to trigger latent positive loops in the system—and off it goes. The difficulty is that once a system is on runaway mode, it is very hard to predict at which next step of equilibrium it will settle. We can see that innovation is the principal source of randomness, but what innovations will bring change? What others will remain inconsequential?

> In the history of humanity, we have recorded a number of such 'revolutions': the first being the Neolithic revolution where the success of agriculture structured human life for the next millennia. The seventeenth century seems to have triggered another series of shocks that we call 'industrial revolutions'—some count four. Each of these shocks seems to push us further in one particular direction but prediction is impossible. Some apparently small changes, such as the introduction of PCs, can have untold consequences. On the other hand, the PC concept is consistent with a consumer society— as opposed to a centralized regime such as the ex-Soviet Union. It is hard to know which innovations will prove critical, but we can expect them to appear in 'small' commercial applications: accessible to mass markets, because that is where our present form of capitalism is going to focus its effort.

Coupled negative loops can create runaway behaviour

Surprisingly enough, the very need for balance and equilibrium of a system can accidentally create positive feedback loops. If we couple negative balancing loops, we inadvertently create a latent positive feedback loop. This positive loop will only trigger runaway behaviour in some particular circumstances—mostly if the pressure on both balancing aspects is roughly equal.

For instance, the escalation process—two elements (persons, departments, companies, nations) getting involved in fights which quickly 'degenerate' while each side only feels that he or she is naturally responding to the 'other's threat'—can only exist if the two elements perceive each other to be of equal stature (if not, one gives). Situations such as present-day Yugoslavia, or Lebanon, can tragically maintain themselves as long as both sides believe that they have a chance to prevail. This escalation mechanism—called *common schismogenesis* in psychology[7]—is particularly difficult to control as one side must 'stop first'.

'Survival of the meanest'—or the fittest—is another such behaviour. Once an element of a system—such as a department in a company—starts to grow (for whatever reason) it will tend to expand so fast that it can outweigh all the other elements in the system. This in itself can prove fatal to the system as a whole which might *need* these elements to function.

Finally, the drift to lower performance: left to itself, a system will tend to follow the law of systems gravity and, by setting its goal on its previous performance, achieve less and less—until it grinds to a halt and desegregates into more dynamic subsystems.

Why competition gets out of hand: escalation

Have you ever wondered how insignificant incidents can escalate into full-blown conflict? In fact, when the first person feels aggressed, he or she will respond in the way he or she knows best; which will then reinforce the other person in his previous threatening—or annoying—behaviour. By referencing our behaviour on each other, we escalate the overall situation.

The most classical example is that of any 'cold war': both countries' military establishments have 'rational' views of the situation:

For country A, the need to build more arms is driven by the threat posed by the other side's existing stockpile of arms (Figure 9.5).

Similarly, country B's generals assess A's military capacity, and

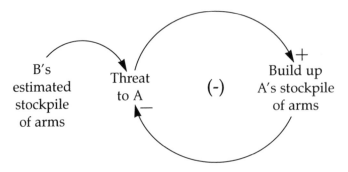

Figure 9.5 A's vision of the situation

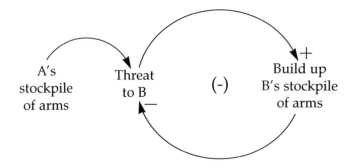

Figure 9.6 B's point of view

respond to this threat by arguing for more weapons (Figure 9.6).

These two balancing processes coupled together lead to an outcome no one wants: the more A builds arms, the more it threatens B, so the more B builds arms, and so forth. Both countries are now stuck in an 'arms race' pattern (Figure 9.7).

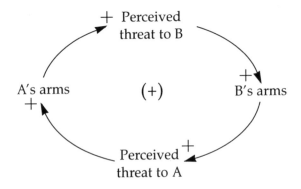

Figure 9.7 The overall outcome: escalation

The ironic—or dramatic—thing about this behaviour is that both sides are building up considerable military capacity that none can use for fear of retaliation and this in itself has been argued as a 'rational' defence strategy by people who have at great cost accumulated enough weapons to destroy the earth several times over.

Each side, however, feels entirely justified and argues that 'If the other side would slow down, they could stop all this nonsense and do some serious works. In business this behaviour is frequent when two competitors of roughly equal size face each other. They can wage war on prices, advertising, staff—or, on the contrary, create a false sense of security by 'agreeing to a status quo' as the American Big Three did until the Japanese competition arrived. Because the world of business is such an open system, escalation tends to lead to the exhaustion of both players—to the advantage of a third, hitherto unknown competitor.

A City bank was concerned about the influences banks had on the British economy. It was felt that rather than acting as a stability factor, the banking sector would in fact emphasize upwards or downwards trends. A systems analysis highlighted the following mechanism: the bank's approach to granting credit was not built on 'objective' analysis of risk, but on competitive risk. Any individual bank would align its lending policy with what its competitors were doing at the time—in order to minimize its competitive exposure. These 'balancing' loops coupled together create a positive feedback structure that is likely to reinforce any changes in perception about the state of the economy and hence the risk of lending (Figure 9.8).

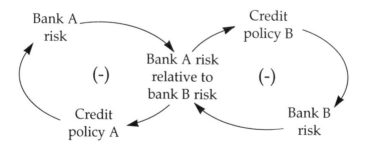

Figure 9.8 Balancing loops create a positive feedback structure

Growth then becomes boom, where the banks are competing against each other to provide more loans, at more competitive rates and therefore they artificially inflate the credit structure of the economy—which finally bursts, and then banks will lead the band-wagon downwards by being increasingly unwilling to lend because they would not want to expose themselves more than their competitors. Such behaviour can only be broken by government intervention (unlikely) or by an 'objective' assessment of risk— irrespective of how much, or how little, lending other banks are doing (highly unlikely).

Survival of the fittest: the dangers of Darwinian policy making

Another case of coupled negative loops creating a positive loop is that of 'the rich get richer and the poor get poorer structure'. If two departments or projects are in competition and create tensions in the system, one might start showing short-term results and then move on to 'win the game'. For example, while production is pulling one way, finance may be pushing the other way, and marketing taking yet another contrary action. What is more, if the tensions are reinforced by a 'all I care about is results' policy in the firm, the instability of the system will usually resolve itself by the 'victory' of one of the actors, in a *success to the successful* type of pattern (Figure 9.9).

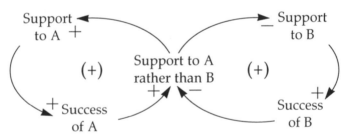

Figure 9.9 A self-fulfilling prophecy

In the case where two activities compete for limited support or resources, success justifies resources: the more successful one becomes, the more support it gains. Consequently, while one of the two interrelated activities is beginning to do very well, the other is having increasing trouble to keep up. In some cases, when both activities are necessary to the overall outcome, this sort of drift can become near lethal.

An American car manufacturer created its own financial institution, originally to support credit policies. This subsidiary soon did very well, and it was increasingly realized that the corporation made more money on the credit to a car buyer than on the purchase price itself. The finance subsidiary diversified into leasing of various transports such as planes and freighters, and did very well indeed; to the point that any budget to leverage financial operations was justified—even if it meant that ailing production would not get it.

Talented executives in the corporation knew where money was to be made in bonuses and such, and they started to drift towards finance (slicker suits too!) rather than stay in the manufacturing side of things. Everyone forgot that in order to sell financing for cars, one has to sell the car in the first place. Fatally, the manufacturing division got in such trouble that in order to generate cash, the corporation had to sell an asset. The only attractive asset they found to sell was their finance subsidiary.

Downhill drift

Organizations which take their objectives from their own performance are likely to deteriorate. These mechanisms particularly afflict bureaucracies. If, for some reason, performance falters, then the new lower performance will be accepted as the new goal and become the standard. Less corrective action will be taken and performance will go down further, and so forth. This can be described by adding another balancing loop to our representation of the organizational feedback loop (Figure 9.10).[8]

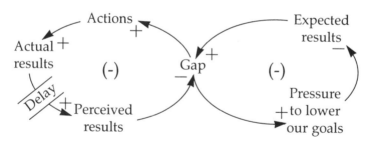

Figure 9.10 Deterioration in organizations

This mechanism frequently appears in administrations *where the overall goal is confused*, corrective actions are extremely rigid and, as a rule, objectives are set by previous performances. The positive loop shown in Figure 9.11 is in place.

An obvious answer to this situation is never to let standards drift,

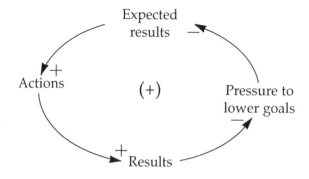

Figure 9.11 The dynamics of decline

even a little or, as often happens, 'just while we get through this crisis' being sure that you will not be able to get back to higher standards later.

> An IT consultancy selling specialized service (complex analysis, statistics, expert systems) at premium price found itself going through a 'bad patch'. The founders were reluctant to shift their positioning and accept to discount, but under pressure from their colleagues, they accepted to do a few 'systems building' jobs for less than their standard rate. Over two years, they found that such contracts were more frequent (demand for low skill/low margin work was considerable) but that the types of contract 'they used to do' were becoming harder to get. Although they were permanently busy, their margins were not half as good, and once or twice they found themselves doing a job at cost price in order to 'get the foot in the door'. The founders wondered what had happened, and whether it was worth working so hard for so little reward—both in fun and £s. In this case, they had let their goals erode and finally found themselves in a situation that did not fit any more with what they intended to do in the first place.

Nothing fails like success

This last mechanism of eroding goal describes a risk complex systems run when they operate in too 'comfortable' an environment: they can over-evolve—that is, adapt so perfectly to their stable environment that they lose any capacity to change and innovate. When a shift occurs, they rapidly become extinct because they have no adaptive capacity left. In a way, they are totally *committed* to their environment. To quote historian Arnold Toynbee:

in evolution [the saying] 'nothing fails like success' is probably always right. A creature which has become perfectly adapted to its environment, an animal whose whole capacity and vital force is concentrated and expanded in succeeding here and now, has nothing left over with which to respond to any radical change. Age by age, it becomes more perfectly economical in the way its entire resources meet exactly its current and customary opportunities. In the end it can do what is necessary to survive without any conscious striving or unadapted movement. It can, therefore beat all competitors in the special field; but equally, should the field change, it must become extinct.[9]

Consultant Richard Pascale advocates the need for *fit* and *split*. Fit is what large organizations need in order not to go to pieces. In systems thinking terms, fit is alignment, consistency around a set of guiding principles and values that will protect the system from fighting against itself. Too much fit, however, leads to over-evolution and fragility to environmental change. This author[10] argues that many of the so-called 'excellent' companies identified by Peters and Waterman in *In Search of Excellence* have now fallen from their pedestal due to an overdose of fit.

To compensate for fit, he argues that large organizations must develop some 'split'. The idea is to break things down into pieces when things get too big or too homogeneous. Split involves maintaining conflict as a transformational force, a source of new ideas and creativity. Constant innovation, for instance, is a necessary provider of split.

Variety: survival insurance

Variety is a necessary condition of survival for complex systems. Only through variety can the system ensure that if one part breaks down with a shift of circumstances, it will survive overall because another—unexpected—element will take over. Unfortunately, variety tends to be costly, and most human organizations tend to reduce variety rather than increase it. The reason behind that attitude is that variety inevitably brings conflict; and conflict is usually seen as negative. Senge suggests that:

> Contrary to popular myth, great teams are not characterised by an absence of conflict. On the contrary, in my experience, one of the most reliable indicators of a team that is continually learning is the visible conflict of ideas. In great teams, conflict becomes productive ... On the other hand, in mediocre teams, one of two conditions usually surround

conflict. Either, there is an appearance of no conflict in the surface, or there is rigid polarization. In the 'smooth surface' teams, members believe that they must suppress their conflicting views in order to maintain the team—if each person spoke her or his mind, the team would be torn apart by irreconcilable differences. The polarised team is one where managers 'speak out', but conflicting views are deeply entrenched. Everyone knows where everyone else stands, and there is little movement.[11]

Not surprisingly, such mediocre teams break down under pressure. When the going gets rough, the players do not know how to deal with their inevitable differences—and the whole affair goes to pot.

Conflict of ideas rather than conflict of people

In many organizations, 'conflict' is a bad word. However, there is nothing inherently wrong with conflict itself—only with its expressions.

> When talking with employees of automakers in different parts of the world, researchers Hervey and Richardson[12] noticed that some said they avoided conflict whenever possible in product development discussions, while others seemed to thrive on conflict in much the same circumstances. To confuse them even further, those who seemed 'positive' about conflict seemed to perform better in many respects in their product development (and overall organizational performance) than those who discussed conflict 'negatively'.
>
> Probing further, they found that different people (and organizations) used the term conflict in very different ways. For some, those who perceived conflict as negative, conflict was personalized and represented destructive personal tensions. For others, those who viewed conflict as positive, conflict was simply an intellectual disagreement to be resolved and had no relation at all with their feelings about the other party. The first approach can be called conflict among people (or destructive conflict) and the latter, conflict of ideas (or constructive conflict).

Conflict of ideas—as 'split'—is a necessary condition for the survival of any human system. Rather than trying to coerce people into agreement, to lower the global level of 'conflict' we have to learn to treat each other with the 'benefit of the doubt. Ideas are not real, the map is not the territory, the word cat will not scratch. Variety and creativity—which means conflict of ideas—are our only guarantees of survival.

When growth occurs, subsystems appear

Of course, an overdose of split is not likely to help things either. When growth occurs past a certain point, subsystems will appear. These different units tend then to fight each other. The difficulty is to keep the control at the lowest level for flexibility while maintaining alignment. In systems terms, Draper Kauffman[13] describes this behaviour as 'The tragedy of the commons':

> Tragedy of the commons is a structure that derives its name from an essay of that name by ecologist Garett Hardin. In medieval England, most villages had a 'common pasture'. This pasture was available to all and everyone could keep their cattle there. Problems start if a positive loop is created by the feeling that 'the more cows I have, the better off I am. Feeding them is free so I can increase my herd as fast as possible.'
>
> Free lunches, however, are uncommon. This attitude creates a situation which each individual is powerless to avoid. As each person increases his or her herd, the number of cows on the common increases. Past a certain point, the cows eat the common grass faster than it can grow—and they start depleting the common resource. After a while the cows are all starving and the entire village is faced with starvation.
>
> This apparently simplistic fable describes to the letter the tragic situation of Sahel peasants who were turning their land into a desert. Civil engineer Anthony Picardi showed how the desertification of the Sahel was led by villagers using international aid funds to increase their herds—on a land that could not sustain them.[14]

In this structure, individuals use a commonly available but limited resource solely on the basis of individual need. At first they are rewarded for using it; eventually, the resource is either significantly depleted, eroded, or entirely used up. It can be represented by the influence diagram comprising Figure 9.12.

The difficulty is to regulate such a system without limiting the freedom of the subsystems. In that light, we can consider giving out 'principles' from the centre while leaving latitude in the form of the application.

> A large electrical parts supplier was successfully managed on the principle of a reduced corporate centre and maximum autonomy to the branches. This, however, led to instances where branches were competing against each other—particularly in the purchasing area. As a result of a competitive drive to purchase materials at the

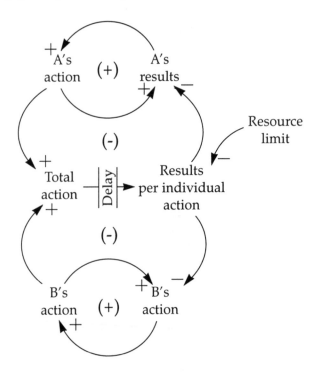

Figure 9.12 The tragedy of the commons

cheaper price, the branches were losing out as a whole because they would inflate prices by out-bidding each other.

Consultants were called in to resolve the situation and they put forward a proposition to create a centralized buying function at corporate level which would take care of purchasing for the branches. This solution was received with scepticism by the MD because it went against all the tenets of the firm's organizational strategy. Finally, the firm settled on a 'cooperation' scheme where branches were given incentives to get together—on their own account—to purchase jointly where appropriate.

Overview

Past a certain size, and a certain degree of complexity, systems seem to take a life of their own. This chapter examines in more detail how change actually occurs in large and complex systems. Although most complex systems tend to be self-balancing and self-organizing, some control loops can inadvertently be coupled and generate explosive growth of one element of the system. This has implications for evolution and adaptability of systems—and also in terms of organizational 'unexpected disasters'.

Notes

1. W. Ashby, *Design for a Brain*, 1952, John Wiley & Sons.
2. G. Bateson, *Steps to an Ecology of Mind*, 1972, Ballantine Books.
3. A. Rosenbluth, N. Wiener and J. Bigelow, 'Behaviour, purpose and teleology', *Philosophy of science*, **10**, 18–24 (1943).
4. J. Lovelock, *Gaia: A New Look at Life on Earth*, 1979, Oxford University Press.
5. For a more detailed discussion of dynamic equilibrium see J. de Rosnay, *Le Macroscope*, 1973, Editions du Seuil.
6. I. Prigogine and I. Stengers, *La nouvelle aliance*, 1979, Gallimard.
7. See G. Bateson, *Steps to an Ecology of Mind*, 1972, Ballantine Books.
8. See D. Meadows, 'Whole earth models and systems', *The CoEvolution Quarterly*, 1980.
9. In R. Pascale, *Managing on the Edge*, 1990, Simon & Schuster.
10. Ibid.
11. P. Senge, *The Fifth Discipline: The Art and Practice of the Learning Organisation*, 1990, Bantam Doubleday.
12. R. Hervey and B. Richardson, *Understanding Information, Communication, Decisions and Learning: Key Insights for Improving Automotive Management*, 1993, Sigma Associates.
13. D. Kauffman, *Systems 1: An Introduction to Systems Thinking*, 1980, Future Systems Inc.
14. In D. Meadows and J. Robinson, *The Electronic Oracle*, 1985, John Wiley & Sons.

Chapter 10

Bigger, better, faster, more!

Are we addicted to growth?

In this day and age, growth seems to be about the only thing people really care about. To be fair, economic growth conditions are so much of everyone's life that this obsession with growth is understandable. It is not, however, linked to 'human nature'. Some societies which have been painfully aware that they were close to overstraining their resources—such as pre-empire India—have focused their social systems on non-growth. The caste system in India, for instance, is built to make sure that 'everything' stays more or less the same in an overcrowded and under-resourced environment. Anthropological studies of the American Indians have shown similar behaviour with, for instance, the Navajo attitude towards wealth. A rich man is looked down upon for if he is rich 'he doesn't take care of his relatives'—the ultimate Navajo social *faux pas*. Our society, however, has been built on expanding frontiers where—literally—the sky is the limit, and therefore growth is good.

As we can expect, such an attitude is reflected in the very structure of the business world. For instance, the concept of capitalism is based on a simple idea: I lend you some money (capital), so you can make it fructify through thrift and industry (profit) and pay me back bit by bit (dividend). Now, in order to make the capital 'fructify' more we found that investment was a good idea, because it leads to bigger assets, and usually large profits at less risk. In this way the money left from paying back the investor is used to increase one's assets by capital expenditure (Figure 10.1).

This mechanism is reinforced when some clever people invent financial leverage and stocks. Now, the larger the investment, the more attractive the expected returns, which generates leverage and a capacity to invest. Furthermore, the larger the profits, the larger the dividends—or the more reliable. The better the share price the more leverage, and so on. All very well, but we are dealing with positive

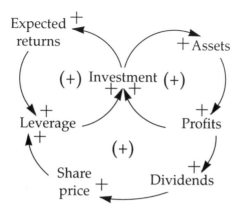

Figure 10.1 Addiction to growth: the rule of money

loops: as long as everything grows we are happy and progress exponentially, but if something snags, then the entire structure is likely to collapse.

Furthermore, we must keep in sight that most of this 'wealth' is paper money. It has little relation to actual physical wealth and is only a collective abstraction. Consequently, the value of this paper wealth tends to grow 'by itself'—through competitive dynamics that inflate the numbers. The entire system works as long *as it grows*. When it stops growing, many of these vested interests are found to be absurd, which can cause major economic and social traumas because of our incapacity to separate the measurement system— paper money—and the actual wealth produced. We are in no way arguing against money, nor suggesting we should go back to bartering as an exchange mode, simply pointing out that the more sophisticated the system becomes, the more it is dependent on growth for its well-being. We are, therefore, addicted to growth: we cannot afford not to grow, because it could mean our demise.

What makes it work: economies of scale

For this reasoning to function, we need to be able to transform larger assets into more profit. In many ways industrial behaviour is still conditioned by Ford's discoveries: by standardizing components and increasing volume the cost of production can be dramatically reduced. This fundamental insight has contributed largely to the success of American industry. The original idea has been adapted, refined, modified and yet remains at the heart of our understanding of mass production. However, nothing fails like success. Any policy—no matter how wise—can become obsolete and even dangerous if it is turned into gospel. Henry Ford himself was confronted by this problem. The logical implication of his fundamental insight generated, at the extreme, his stand on 'customers can have any car they want as long as it's a black model T'. People wanted choice as well, and when Alfred Sloan applied marketing concepts to the car industry, GM stole the day. This did not invalidate the advantage of economies of scale but simply showed that if low cost through standardized high volume was necessary, it was not always sufficient in itself, and as a dogma could lead to disaster.

> If we take a broad view of the problem, we can split manufacturing costs into two categories: those which respond to volume or scale and those which are driven by variety. Scale-related costs should decline as volume increases, usually falling by 15 to 25 per cent per unit each time volume doubles. On the other hand, variety-related costs reflect the cost of complexity in manufacturing: as variety increases, cost increases at a rate of about 20 per cent per unit each time variety doubles. For that reason, the temptation in manufacturing is to limit variety and to increase volume.

Self-constrained growth

Because growth is necessary to keep the system running, most small companies will sooner or later go for growth in a big way. As it happens, their growing action often backfires—which is often known as the 'five years crisis'. Rather than blaming the economy or bad luck, a systems analysis shows how companies can be limiting their own growth as a result of their own policies. In the growth of any organism, the only important factor at any one time is the most limiting. As a grain of sand can jam an entire mechanism, the most limiting factor can stunt growth very rapidly. However, because this most limiting factor is seldom the largest one, it tends to be totally ignored—until too late. For many young companies, the most limiting factor tends to be their clients' goodwill. In many cases, as the company grows quality slips—whether in the actual product or in the service area such as delivery time or maintenance—and as quality slips, clients simply do not come back. When the same thing happens with a large company its leaders usually have more time to correct the situation, but with a small company the tyranny of cash

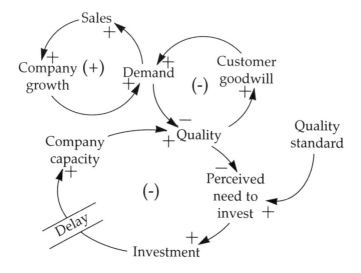

Figure 10.2 Self-constrained growth

flow can dramatically accelerate events. Jay Forrester[1] has modelled this structure as shown in Figure 10.2.

As demand increases—generally pushed by sales, and plenty of 'nice' positive loops—the system tends to be strained and the quality of the firm's performance slips. Growth then approaches a limit which could be pushed into the future, or even eliminated, by investment in additional capacity. This investment, however, needs to be aggressive in order to forestall reduced growth, or else it will never be made. If the firm waits too long, it might find that key goals or performance standards are allowed to slip in order to justify under-investment. This leads to a self-fulfilling prophecy where lower goals generate lower expectations, which are then justified by poor performance caused by under-investment.

A high-tech manufacturer of high-powered work stations had an astoundingly successful first year. Production was strained at its maximum, but still more orders were coming in. As a result, the lead time on orders were growing up to one or two years. Customers faced with such lead times were dissatisfied and cancelled their orders. Soon the word got around that the company was unreliable and could not deliver, and more orders were cancelled. In the face of this reversal of 'luck' the owners felt that it would not be safe to invest in additional capacity 'before things got better'. In the words of the MD: 'Well, we used to be the best, and we'll be the best again, but right now we have to conserve our resources and not over-invest.'

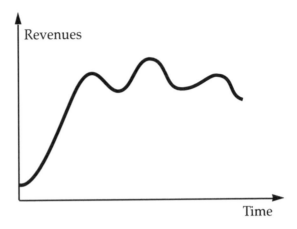

Figure 10.3 A strained system

The performance of such a firm will tend to look like Figure 10.3.

The investment dilemma: make room for growth

The main difficulties growing companies face is to balance out the risk of investment—which usually implies debt—and the need for enough capacity to have 'room to grow'. We can see that the difficulties a company's owners face are then stuck between a rock and hard place: either they wait until orders come through in order to invest safely—and then lose quality and clients because of the delays involved with turning investment into up-to-speed production capacity—or they invest in additional capacity beforehand and run the risk of not being able to pay back the bank if the orders do not arrive.

In most cases, the pressure will be for growth. Most investors expect rapid growth from new companies, the difficulty is how to sustain this growth and not to eat up budgets before the income is there. A good rule of thumb is to try to set one's growth rate on one's current profitability. The error to avoid is to believe that profitability will immediately increase with expansion. It should, eventually, but in most cases profitability *falls* with expansion. The unexpected costs of growth are always larger than we anticipate and the delays involved in getting new capacity 'up to speed' are greater.

The key element in investment capacity is delays. If we could invest and the next day have our additional capacity up to speed with the rest of the operation we would be fairly secure in our investment decisions. The snag is that in most companies structure follows the decision and not the other way round. For instance, if we decide to hire more we are going to argue about the numbers for months, finally a project comes in and we need three more people working on it tomorrow. Typically, we hire in a rush, then realize our offices are already overcrowded. So we rent out the first available offices which are impractical and expensive and which takes three months. Meanwhile, back at the ranch it is chaos and confusion. Then it appears that these people we hired are not quite up to speed so the old hands have to do some of their work—on top of their own already out-of-control workloads. In a few weeks complaints arrive about the wages these new people are earning which are far superior to what some of the previous employees get, and so forth. Some companies cope, many fold: the new contract is a failure, clients are dissatisfied, there is no money to pay this increase in capacity.

The issue is that most of the structural elements such as larger offices, a pay and promotion plan, a training schedule to get new people up to speed could have been prepared beforehand—at low cost. In effect, one can make room to grow by setting up an environment that can accommodate a larger activity. Most companies will not go to the trouble because they feel that they are already too busy to do anything 'bureaucratic' and then they feel that if they are small they do not need such heavy structures. The issue is that the structures, as the premises or land to build a new plant, need not be activated until it is necessary.

> Jay Forrester recalls the early days of Digital Equipment Corporation. In those days, a few of his former students started the company from a corner of one floor in an old mill building outside Boston. As a member of the board, Forrester persuaded them to rent the whole football-field-size floor as soon as the space became available. That leap in capacity–which at the time seemed outrageous–allowed Digital to grow without being constrained by the most trivial limit: premises. A most dramatic experience, Forrester said later, was to come back only six months later to find the entire floor full of people working hard at it.[2]

In making room for growth, we are not over-investing, simply making sure that we can grow *when and if we need it*—and not six months after the battle. In this way we can avoid being stuck in the cycles of the self-limiting growth.

Phases of growth

Several authors studying business growth identify different phases. These phases tend to be those of: creation, survival, growth, expansion and maturity. These phases can be best understood by looking for the most limiting factor—and how it was created by the solution found to the problems of the previous phase.

Creation

For instance, at the creation phase the growth engine tends to be expertise, and the limiting factors financial administrative and selling skills.

> A prestigious technical university developed its own space programme: the idea was simple, to manufacture and launch 'small' commercial satellites. Rather than building the expensive complex satellites traditionally used, this project adopted a 'minimalist' approach, and built small, simple and cheap satellites. These satellites appealed to commercial users who had a need for satellite transmission, but could not afford standard rates. The university project became a commercial operations.
>
> The founders were either professors or ex-students. By and large, they scorned anything associated with management and focused on building the 'best' satellites. They assumed that the product was right and that it would 'sell itself'. Members were not interested in money; they spent long hours experimenting with the satellites, were very flexible and relied considerably on the originator of the project who seemed to 'know everything'. The project was somewhat secure inasmuch as it was still largely funded by the university.
>
> However, in the first year, sales were not as high as expected. Furthermore, they had trouble keeping to their deadlines because of development delays they could not control. Administration was a nightmare, particularly after the secretary who kept it all in hand had a flaming row with the project leader. Two years or so down the road, confidence is not what is used to be and some ex-students are starting to review their career choices.

Such a situation can usually be remedied by training in the financial, administrative and selling areas. The founders need to accept that they will have to devote more time to 'management' issues. An effective marketing campaign can turn up trumps.

Survival

From the creation phase, the firm moves into the survival stage where managerial capacity is usually the critical limiting factor. If training is not sufficient, the business needs to consider appointing a strong business manager.

Growth

Survival is generally followed by growth. This usually means a strong structure and functional departments. The limits will now be around administrative behaviours, and capacity overload. Delegation, running the business as a team and a robust financial control and understanding that allows for capacity investment should overcome this phase.

Expansion

At this stage, expansionism is running the business. This usually poses autonomy problems, staff morale problems and a rise in overheads. We might consider breaking down the company into smaller units; securing a long-term debt and considering new partners.

Maturity

Eventually the main product becomes dated, the company is diversifying horizontally and shareholders are putting on the pressure. Innovation and improved productivity can protect the organization. At this stage, it is critical to maintain both shared values and diversity in order to be able to survive.

Sustainable growth

As we focus on growth of individual companies, we tend to forget the wider issue of growth at a social level. Fundamentally, our organizations are 'open systems' which is a neat way of saying that for the machine to work it must pump resources from 'outside' and throw back waste 'out there'. As we have seen, growth is more than an obsession: it is an addiction. The individual firm cannot afford to remain static in the present dynamics of the marketplace. This is a sad state of affairs, but a fact of life. We can claim with E.F. Schumacher that 'small is beautiful', but the fact remains that 'big is powerful'.

As a result, much in the style of the 'tragedy of the commons' on a grand scale, individual firms push for growth at all costs in order to survive in their competition against their neighbours, and in doing so, the entire human race is accumulating a sinister bill which—much like the budget deficit—future generations will have to pay. The question is: 'is sustainable growth possible?' Can we grow without depleting our most needed resources? Or, in other words, can we grow intelligently? The authors of the original book, *The Limits to Growth*, claimed that we were driving towards a cliff without headlines on the basis of a systems dynamics computer simulation, state the problem thus:[3]

> [T]he three conclusions we drew in *The Limits to Growth* are still valid [20 years later], but they need to be strengthened. Now we would write them this way:
>
> 1. Human use of many essential resources and generation of many kinds of pollutants have already surpassed rates that are physically sustainable. Without significant reductions in material and energy flows, there will be in the coming decades an uncontrolled decline in per capita food output, energy use, and industrial production.
> 2. This decline is not inevitable. To avoid it two changes are necessary. The first is a comprehensive revision of policies and practices that perpetuate growth in material consumption and in population. The second is a rapid, drastic increase in the efficiency with which materials and energy are used.
> 3. A sustainable society is still technically and economically possible. It could be much more desirable than a society that tries to solve its problem by constant expansion. The transition to a sustainable society requires a careful balance between long-term and short-term goals and an emphasis on sufficiency, equity, and quality of life rather than quantity of output. It requires more than productivity and more than technology; it also requires maturity, compassion, and wisdom.[4]

Is more better?

Yet, if we are caught in the dynamic, how can we act without waiting for an all-powerful, benevolent government to do it for us? How can individual firms contribute to a sustainable society while staying in business and profitable? The first thing to realize is that the two are not necessarily contradictory. For instance, most of the American firms that have been forced to 'clean up their act' to follow government pollution legislation have discovered significant productivity increases linked to new, cleaner technologies. At the root of the problem is our 'addiction to growth'. This mechanism, however, contains a number of logical blind spots: the most obvious one is that it rests entirely on the idea that more is better. More is better is a concept profoundly anchored in our thinking—and our day-to-day experience—but not always necessarily true. This idea that there is not enough to get by and so we will always need more rests on (1) a fundamental paranoia and (2) a disregard of the notion of value.

More is not better, but more value is. Somewhere along the line we are confusing quantity and quality. For instance, would you prefer ten average television sets or one really neat one? The confusion comes from our difficulty to distinguish between *the thing* and *the use* we make of it. The thing is easy to understand: it is right there in front of you, it can be counted, etc. The use is a rather more relative and abstract concept; yet it is what matters. Things in themselves rarely have any value (apart from works of art)—who would be crazy enough to own a washing machine that does not work, or keep washing powder on the library shelf? The value of the washing machine and the washing powder is in their ability to wash, not in the things themselves.

> As a result of an experiment which went wrong, the industrial giant 3M invented a glue that would not glue. It was sticky at most, but certainly would not hold anything together. However, someone had a brilliant idea and applied this glue to little pieces of paper—creating the Post-it note which now covers most working areas. The value of a product is in the benefit it provides to its customer—a truism, yet so often forgotten by manufacturers who tend to value products in themselves.

To be fair, this confusion comes from an in-built capacity of the human mind to attribute value to objects, symbols, actions, etc. Suppose I pick up a stick in the woods and place it on the mantelpiece. Every morning, I come down and stop to look at this stick. Very soon, the stick will have a certain value for me. Before I know it, I will find myself praying to the stick.

> The old master of a Zen monastery suddenly felt his end near. He looked around him and could not think of whom to appoint head of the monastery to succeed him. One morning he placed a jar on a chair, and asked each and every inhabitant of the monastery to come and make a speech to the jar. As the monks took their turns he heard all possible manner of speeches—studied, scholastic, witty, eloquent—and yet was not satisfied. At the very end of the line, the cook, a genial and uneducated man, looked at the jar, frowned and picked it up. When the master asked him what he was going to do with it, the cook smiled and answered, unconcerned

'put it back where it belongs, on the kitchen shelf'. This man was then chosen to become the next head of the monastery.

We often attribute some value to objects, actions or attitudes not so much for the benefits they bring but for social reasons. Mechanisms such as 'keeping up with the Jones's', promoting an idea because it is the 'in' thing to do at the moment or accepting to buy over-priced goods at their 'market value' are a structural aspect of collective life and action. We attribute value to things when we see that enough people are already doing so and it fits within our preoccupation. The issue is that 'fashionable' solutions do not necessarily bring the benefits they advertise. Rather than take the situation out of context, we must focus on the entire context and evaluate all the costs and benefits in order to find the 'real' value of our action. Sometimes, our answers will be unfashionable ... so what?

Value and efficiency

If we pursue value—not quantity—then we must look into the question of efficiency. Efficiency is the ratio between outputs and inputs. To be efficient is to get more for less; yet as we can see, there is a considerable difference between quantity efficiency and quality efficiency. Technology tends to be a vector of efficiency—higher technology usually means a source of efficiency. However, there is nothing in technology itself which says that it will be applied to qualitative efficiency rather than quantitative efficiency. Because of this confusion and our belief that more is better, we tend to apply technological improvement to the wrong problems.

In the 1980s, many theorists of the 'Leisure Society' made the following reasoning:

1. Technology makes things more efficient.
2. Therefore, we need fewer people to do the same amount of work.
3. Consequently, if we keep the same number of people employed, we will need to make them work less for the same level of output: we will be qualitatively more efficient.
4. This leaves room for a leisure society—in which people will have time to do what they really want to do.

As it turned out, we have solved the equation differently. Because more is better, we keep fewer people employed, working longer hours and the rest go unemployed. It is an old story. Palaeontologists now think that hunter-gatherers would spend about two to three hours a day gathering food. After the Neolithic revolution, a peasant would spend most of his day working the land!

Strategic consistency

There are no right or wrong strategies: only consistent or inconsistent ones. Corporate strategists often argue about the 'right' strategy to follow. 'Go for market share!' or 'focus on your margins' or 'centralize your operations' and 'decentralize your organization', they cry. Most of these injunctions appear impossible to follow all at once—yet some of them must be right, so how shall we choose? However, there are no 'right' or 'wrong' strategies. Success or failure is determined by an action that is consistent with the outcomes we want to see happen, and the shape we are in to start with.

Companies do not exist in isolation, without a market or a past. In most cases, they have evolved in a particular form that corresponds to the initial context of their activity. In this context, they run very well, they have learnt to be very effective at what they do. The problem is that this context can change quite dramatically, sometimes as a result of their very action—as in the case of a firm that develops a successful new product and opens up a market for other competitors.

Apple computers is a striking example of that situation. After carving a profitable niche in personal computers, the company is now facing hard times. Regardless of specific strategic choices on new technology, Apple faces a true dilemma: its core business, selling Macintosh computers, has become uncompetitive. Since its beginnings in 1984, the Mac has transformed computing by making it user-friendly and widening the possible applications. Originally Mac users were happy to pay premium price for such benefits. But as the success of the 'Mac concept' grew, other players joined the field. According to The Economist[5] some 27 million PCs have been equipped with Windows since 1990—compared to the 10 million Macs sold since 1984. With little to differentiate it (The Macintosh is still the 'real thing' to any Mac lover) Apple has been forced to cut prices and fight it out with the PC clones. However, it is a losing battle because Apple's raison d'être—a desire to be different—is fundamentally inconsistent with low cost manufacturing.

The firm must now 'change'; it must adopt a new 'business strategy'.

We can easily say that firm A must learn to produce at low cost, or that firm B must become more creative, etc., but companies retain what sociologist Catherine Ballé calls their 'institutional heritage'—sets of values, rules and habits that make the company what it is. Some can be changed, but some others are so deeply ingrained that they cannot be changed. Yet, this institutional heritage is also a great source of wealth if it is not ignored. In order to adapt, an organization must delve into its very matrix and identify the few possible routes it can take, to find its own way rather than follow some trendy generalization about 'how to compete in the 1990s'.

Growth itself can be seen as a stochastic process built upon growth engines and limiting factors. Imagine a child's toy—a frame with a few holes of different shapes: a round hole, a rectangular hole, a square hole, a star-shaped hole, etc. Through this frame, we are going to sift some sort of blocks made of putty. These blocks have an original shape, but they are flexible enough so that we can modify them slightly if we need to. The only elements that will go through are those which contrive to have the 'right' sort of shape: they do not all need to be equal, but they need to be consistent—half a star and half a square probably will not work.

Systems thinking shows that certain organizational strategies will create their own limitation, and the corporation will need consequently to shape itself according to these imperatives.

> For instance, if you go for explosive headcount growth, you will need to focus on a strong structure and keep an eye on culture and quality. However, if you stay small, and push for profitability on expert products, you will need to maintain your creativity at all costs—which probably means individualism and chaotic management by trial and error.

Many companies grow simply to administrate the success of a product. Because they get rich and fat through luck or brilliant insight that happened a long time ago, they tend to assume that their working practices are sound (*we must be doing something right*). However, when they hit an unexpected snag, these cultures do not know how to adapt and the consequence is often a sharp downsizing—painful and destructive. Unless corporations learn about their growth and the limits that they will face, they cannot adapt.

How to build a new approach to strategy

Strategy is about planning for the future. It has traditionally been seen as defining objectives, and then building operational goals from there, measuring the gaps between results and these goals, and trying to close these gaps. Also strategies have consistently succeeded and failed in a somewhat random fashion. In fact, most 'strategic thinking' seems to assume a 'clean slate', a Cartesian *tabula rasa*. The organization is seen as a body of 'resources' or 'competencies' but its current actions and policies tend to be dismissed. However, these policies will continue to have an effect while we devise the 'strategy'. Life goes on, the dynamics that we see now will go on. We can modify some of it, channel them but not consider that they do not exist. And as time passes, the very fact that these dynamics exist changes our environment so that we will need to change them.

> The qualities which make a child successful with his or her peers and parents are very different from those which he or she will need to affirm himself or herself as a teenager, which are different again from those needed to succeed as an adult, then a parent, etc. How many teenage stars do we all know who disappeared in college years? It is the same person, but if a person stops learning, then he or she will lose the potential for success. The same is equally valid for organizations—they have to keep learning, to realize that what limits their growth now is different from whatever they had to deal with yesterday, and different again from the limits which will condition tomorrow.

Strategic thinking should focus on the relationship the firm is presently creating with its environment—and the next limiting factor. How is growth itself going to modify these factors? This limiting factor is not necessarily where and what we expect it to be. It can be very mundane, as opposed to strategic. Yet, there lies the difference between self-limiting or self-sustaining growth. Insight is rarely about providing the 'right answer', but about asking the right questions. The questions we ask determine the answers we look for.

To create the structures for growth, in a quality sense, we must focus on the outcomes we want to create and the outcomes our present organization is *actually* creating. The difference between these two can be expressed as *challenges* we will have to face. From then on, what can we modify in our structure to create the outcomes we want? To design a strategy we must ask ourselves:

- What are the outcomes that we want to create?
- What are the outcomes that we are currently creating?
- What are the challenges we face?
- What major constraints can keep us from meeting these challenges?
- If there was one thing to change about how we operate, what would it be?

The objective of growth is not to reach 'the final state' at which we can at last retire—and nothing will change any more. This, we know, means death. There is no ultimate solution. The objective of growth is to maintain the dynamic equilibrium which means life. This equilibrium is forever changing and as it does, we need to adapt, to learn; yet, as opposed to most biological species, this adaptation process need not be random for humans or their institutions. Humans can learn. Humans can control this very adaptive process, if they can control themselves.

Overview

This last chapter examines the implications of systems theory for strategy and corporate growth. It points out some dangerous inconsistencies in our understanding of growth and shows how companies can themselves limit their growth. We then explore how to sustain growth and suggest an alternative handling of strategy making.

Notes

1. J. Forrester, *Collected Papers of Jay Forrester*, 1975, MIT Press.
2. P. Senge, *The Fifth Discipline: The Art and Practice of the Learning Organisation*, 1990, Bantam Doubleday.
3. D. Meadows, D. Meadows, J. Randeu and W. W. Behrens III, *The Limits to Growth*, 1972, University Books.
4. D. Meadows, D. Meadows and J. Randers, *Beyond the Limits*, 1992, Earthscan Publications.
5. 'Apple computers: falling to earth', *The Economist*, August 1993.

Conclusion

Three medieval monks were discussing the nature of causality. They came upon a flag which unfurled in the wind. 'The flag is moving' states the first monk. 'The wind is moving' retorts the second. 'Ahh … the mind is moving' says the third.

The only thing we know about reality is that it is infinitely complex. There is always more to it than we can know—or surely, that we need to know in order to operate in the world. However, because of the uncertainty of that fact, because of the constant surprises of life, we seek stability of the mind. We demand to find some sort of anchor, some sort of resting place, some kind of relationship where there is no longer conflict and change, some kind of knowledge which would be lasting and assured. So, we invent ideology and dogma, any sort of organized belief that will give us deep, satisfying hope. Still, as we can constantly experience, organized belief does not deliver. It divides people, generates absurd actions—in the name of what?

We believe that any attempt to find some kind of reality which will be completely true is bound to fail. Any invention of the mind which will give a significance to our petty problems and our struggles of everyday life is bound to divide and mislead. The map is not the territory. Unless we accept fundamentally that we do not need 'the ultimate map', we will not graduate out of this confusion and division. Maps are extremely useful, but there is nothing sacred about them. Why not make our own as we go?

Systems thinking provides a different method to make maps. It does not reject the validity of other maps: it merely adds to our map-making capability. Sometimes it is more accurate, sometimes less than more traditional approaches. This is what we need our judgement for. We find it interesting because it provides insights into problems which were previously obscure and 'illogical'. However, there are plenty of unsolved questions left to explore.

Why should we bother? Often we feel that the organization, the company, the group is something external to ourselves—something we endure through necessity but that is not us. In the words of the philosopher J. Krishnamurti:

> Most of us in this confused and brutal world try to carve out a private life of our own, a life in which we can be happy and peaceful and yet live with the things of the world. We seem to think that the daily life we lead, the life of struggle, conflict, pain and sorrow, is something separate from the outer world of misery and confusion. We seem to think the individual, the 'you', is different from the rest of the world ... When you look a little more closely, not at your own life but also at the world, you will see what you are—your daily life, what you think, what you feel— is the external world, the world about you. You are the world, you are the human being that has made this world of utter disorder, the world that is crying helplessly in great sorrow.
>
> In bringing about a radical change in the human being, in you, you are naturally bringing about a radical change in the nature of society.[1]

We can feel that considering the size and complexity of our organizations, of our societies, what we do and what we think is of little import. Yet we must take responsibility—take responsibility for how we think. We must realize that our simplistic assumptions of linear causality, of more is better, or of survival of the fittest (somehow things will be all right for the others) create the organization we live in. And through the passing of our experience, we can learn. We can disprove our mental models, create new ones. Learning is possibly the most challenging aspect of individual—and organizational—life. Learning is our only key to growth in quality, and not simply in quantity.

Note

1. J. Krishnamurti, *Talks with American students*, 1970, Random House.

Further reading

Aguayo, Raphael (1990) *Dr Deming: The Man Who Taught the Japanese About Quality*, Mercury Books, London.

Argyris, Chris (1985) *Strategy, Change and Defensive Routines*, Ballinger, Boston.

Argyris, Chris (1990) *Overcoming Organizational Defences*, Simon & Schuster, Boston.

Argyris, Chris and Schon, Donald (1974) *Theory in Practice*, Jossey-Bass, San Francisco.

Armstrong, Scott J. (1985) *Long-Range Forecasting: From Crystal Ball to Computer*, John Wiley & Sons, New York.

Ashby, W. Ross (1952) *Design for a Brain*, John Wiley & Sons, New York.

Ashby, W. Ross (1956) *Introduction to Cybernetics*, John Wiley & Sons, New York.

Ballé, Catherine (1990) *Sociologie des Organisations*, PUF, Paris.

Ballé, Michael and Jones, Trevor (1993) 'Data Visualisation: a fresh look at segmentation', *The Journal of Database Marketing*, **1**, July, 62–76.

Ballé, Michael and Hervey, Richard (1994) 'Survie dans l'industrie automobile', *Progrès Technique*, ANRT, Paris.

Bateson, Gregory (1972) *Steps to an Ecology of Mind*, Ballantine Books, New York.

Bateson, Gregory (1979) *Mind and Nature*, Wildwood House, London.

Beer, Stafford (1959) *Cybernetics and Management*, English Universities Press, London.

Beer, Stafford (1972) *Brain of the Firm: The Managerial Cybernetics of Organisation*, The Penguin Press, London.

Bode, H. W. (1960) *Feedback—the History of an Idea*. Reprinted in Bellman and Kalaba (1964).

Bohm, David and Edwards, Mark (1991) *Changing Consciousness*, HarperCollins, New York.

Boudon, Raymond (1977) *Effets pervers et ordre social*, PUF, Paris.

Boudon, Raymond (1985) *La place du désordre: critiques des théories du changement social*, PUF, Paris.

Boulding, Kenneth (1956) 'General systems theory—the skeleton of science', *Management Science*, **2**(3), 197–208.

Boulding, Kenneth (1969) *Beyond Economics*, The University of Michigan Press.

Boulding, Kenneth (1971) 'The dodo didn't make it: survival and betterment', *Bulletin of the Atomic Scientists*, 27(2), 19–22.

Caspar, Pierre, Grinda, J. R. and Vaillet, F. (1973) *Creez vous-même votre entreprise*, Les Editions d'Organisation, Paris.

Checkland, Peter (1981) *Systems Thinking, Systems Practice*, John Wiley & Sons, London.

Clavel, Bernard (1989) *Maudits Sauvages*, Albin Michel, Paris.

Cyert, R. M. and March, J. G. (1963) *A Behavioural Theory of the Firm*, Prentice-Hall, Englewood Cliffs, NJ.

de Rosnay, Joel (1975) *Le Macroscope: vers une vision globale*, Seuil, Paris.

Dixon, Norman F. (1976) *On the Psychology of Military Incompetence*, Jonathan Cape, London.

Durand, Daniel (1979) *La Systémique*, PUF, Paris.

Fisher, Roger and Ury, William (1981) *Getting to Yes*, Hutchinson, London.

Flood, Robert L. and Carson, Ewart R. (1988) *Dealing with Complexity: an Introduction to the Theory and Application of Systems Science*, Plenum Press, New York.

Forrester, Jay (1958) 'Industrial dynamics: a major breakthrough for decision makers', *Harvard Business Review* **36**(4), 37–66.

Forrester, Jay (1961) *Industrial Dynamics*, MIT Press, Cambridge.

Forrester, Jay (1975) *Collected Papers of Jay Forrester*, MIT Press, Cambridge.

Fritz, Robert (1984) *The Path of Least Resistance*, Ballantine Books, New York.

Genelot, Dominique (1992) *Manager dans la complexité*, INSEP Editions, Paris.

Gleick, James (1987) *Chaos: Making a New Science*, Viking Penguin, New York.

Goodman, Michael (1992) 'The do's and don'ts of systems thinking on the job', *The Systems Thinker*, **3**(6) August, 5–6.

Hall, Nina (1992) *The New Scientist Guide to Chaos*, Penguin Books, London.

Hampden-Turner, Charles (1990) *Corporate Culture: From Vicious to Virtuous Circles*, Hutchinson Business Books, London.

Hirschman, Albert O. (1977) *The Passions and the Interests*, Princeton University Press, Princeton.

Jacques, Elliot (1989) *Requisite Organisations*, Carson Hall & Co.

Janis, Irving L. (1989) *Crucial Decisions: Leadership in Policymaking and Crisis Management*, Macmillan, New York.

Kauffman, Draper L. (1980) *Systems 1: An Introduction to Systems Thinking*, Future Systems Inc., Minneapolis.

Kim, Daniel H. (1993) *Systems Archetypes: Diagnosing Systemic Issues and Designing High-Leverage Interventions*, Pegasus Communications, Cambridge.

Krishnamurti, Jiddu (1970) *Talks with American Students*, Shambala Publications, Boston.

Leibenstein, H. (1950) 'Bandwagon, snob, and veblen effects in the theory of consumers' demand', *Quarterly Journal of Economics*, **64**, 183–207.

Lewin, Kurt (1951) *Field Theory in Social Science*, Harper and Brothers, New York.

Lovelock, J. E. (1979) *Gaia: A New Look at Life on Earth*, Oxford University Press, Oxford.

Lugan, Jean-Claude (1993) *La systémique sociale*, PUF, Paris.

Lyneis, James (1980) *Corporate Planning and Policy Design, a Systems Dynamics Approach*, MIT Press, Cambridge.

Malthus, Thomas R. (1798) *First Essay on Population*. Reprinted Macmillan 1926, 1966, New York.

March, James G. (1988) *Decisions and Organisations*, Basil Blackwell, Cambridge.

Maxwell, James Clerk (1964) 'On Governors', *Proceedings of the Royal Society of London*, Vol. 16, pp. 270–283. Reprinted in Bellman and Kalaba (1964).

Mayr, Otto (1970) *The Origins of Feed-back Control*, MIT Press, Cambridge.

Meadows, Donella H. (1980) 'Whole earth models and systems', *The CoEvolution Quarterly*.

Meadows, Dennis L. and Meadows, Donella (1974) *Toward Global Equilibrium: Collected Papers*, MIT Press, Cambridge.

Meadows, Donella and Robinson, J. M. (1985) *The Electronic Oracle: Computer Models and Social Decisions*, John Wiley & Sons, New York.

Meadows, Dennis, Meadows, Donella, Randers, J. and Behrens III, W. W. (1972) *The Limits to Growth*, University Books, a Potomac Associates Book.

Meadows, Donnella, Meadows, Dennis and Randers, Jorgen (1992) *Beyond the Limits*, Earthscan Publications, London.

Merton, Robert K. (1948) *Social Theory and Social Structure*, Free Press, New York.

Mintzberg, Henry (1989) *Mintzberg on Management: Inside our Strange World of Organisations*, Macmillan, New York.

Morecroft, John D. (1985) 'Rationality in the analysis of behavioural simulation models', *Management Science*, **31**(7) 900–916.

Peters, T. and Waterman, R. H. (1982) *In Search of Excellence*, Harper and Row.

Piaget, Jean (1968) *Le Structuralisme*, PUF, Paris.

Prigogine, Ilya and Stengers, Isabelle (1984) *Order out of Chaos*, Bantam Books.

Richardson, George P. (1991) *Feed-back Thought in Social Science and Systems Theory*, University of Pennsylvania Press, Philadelphia.

Richardson, Lewis F. (1938) 'The arms race of 1909–13', *Nature*, **142**, 792.

Richmond, Barry (1990) *Ithink User Guide*, High Performance Systems, Hanover.

Saunders, P. T. (1980) *An Introduction to Catastrophe Theory*, Cambridge University Press, Cambridge.

Schumacher, E. F. (1973) *Small is Beautiful: A Study of Economics as if People Mattered*, Blond & Briggs, London.

Senge, Peter M. (1990) *The Fifth Discipline: The Art and Practice of the Learning Organisation*, Doubleday, New York.

Shannon, Claude E. and Weaver, Warren (1949) *The Mathematical Theory of Communication*, University of Illinois Press.

Simon, Herbert A. (1957) *Models of Man*, John Wiley & Sons, New York.

Smith, Adam (1776) *The Wealth of Nations*, Everyman's Library, Random Century, London.

Stacey, Richard D. (1991) *The Chaos Frontier: Creative Strategic Control for Business*, Reed International Books, Oxford.

Thom, René (1980) *Paraboles et catastrophes*, Il Saggiatore, Milan.

von Bertalanffy, Ludwig (1968) *General Systems Theory, Foundations, Developments, Applications*, George Braziller, New York.

Wiener, Norbert (1950) *The Human Use of Human Beings*, Houghton Mifflin, Garden City.

Wolstenholme, Eric F. (1990) *Systems Enquiry: A System Dynamics Approach*, John Wiley & Sons, London.

Womack, James P., Jones, Aniel T. and Roos, Daniel (1990) *The Machine that Changed the World*, Collier Macmillan, New York.

Index